Data Analytics

Data Analytics Applications

Series Editor:
Jay Liebowitz

PUBLISHED

Big Data in the Arts and Humanities: Theory and Practice
Giovanni Schiuma and Daniela Carlucci
ISBN 978-1-4987-6585-5

Data Analytics Applications in Education
Jan Vanthienen and Kristoff De Witte
ISBN: 978-1-4987-6927-3

Data Analytics Applications in Latin America and Emerging
Economies
Eduardo Rodriguez
ISBN: 978-1-4987-6276-2

Data Analytics for Smart Cities
Amir Alavi and William G. Buttlar
ISBN 978-1-138-30877-0

Data-Driven Law: Data Analytics and the New Legal Services
Edward J. Walters
ISBN 978-1-4987-6665-4

Intuition, Trust, and Analytics
Jay Liebowitz, Joanna Paliszkiewicz, and Jerzy Gołuchowski
ISBN: 978-1-138-71912-5

Research Analytics: Boosting University Productivity and
Competitiveness through Scientometrics
Francisco J. Cantú-Ortiz
ISBN: 978-1-4987-6126-0

Sport Business Analytics: Using Data to Increase Revenue and
Improve Operational Efficiency
C. Keith Harrison and Scott Bukstein
ISBN: 978-1-4987-8542-6

Data Analytics
Effective Methods for Presenting Results

Edited by
Subhashish Samaddar and Satish Nargundkar

CRC Press
Taylor & Francis Group
Boca Raton London New York

CRC Press is an imprint of the
Taylor & Francis Group, an **informa** business

AN AUERBACH BOOK

CRC Press
Taylor & Francis Group
6000 Broken Sound Parkway NW, Suite 300
Boca Raton, FL 33487-2742

© 2019 by Taylor & Francis Group, LLC
CRC Press is an imprint of Taylor & Francis Group, an Informa business

No claim to original U.S. Government works

Printed on acid-free paper

International Standard Book Number-13: 978-1-138-03548-5 (Hardback)

Library of Congress Cataloging-in-Publication Data

Names: Samaddar, Subhashish, author. | Nargundkar, Satish, author.
Title: Data analytics: effective methods for presenting results / Subhashish Samaddar, Satish Nargundkar.
Description: Boca Raton, FL : CRC Press/Taylor & Francis Group, 2019. | Includes bibliographical references and index.
Identifiers: LCCN 2018061489 (print) | LCCN 2019001236 (ebook) | ISBN 9781315267555 (e) | ISBN 9781138035485 (hb : acid-free paper)
Subjects: LCSH: Business—Data processing. | Business requirements analysis. | Business analysts.
Classification: LCC HD30.2 (ebook) | LCC HD30.2 .S258 2019 (print) | DDC 658.4/52—dc23
LC record available at https://lccn.loc.gov/2018061489

Visit the Taylor & Francis Web site at
http://www.taylorandfrancis.com

and the CRC Press Web site at
http://www.crcpress.com

Contents

Preface

If you play any role in the field of analytics, you can benefit from this book. If you receive the results of someone else's analytics work to help with your decision-making, then this book is for you, too. In the two decades of our involvement with teaching and consulting in the field of analytics, we have witnessed it become an integral part of supporting business strategy in many organizations. Over time, we noticed a critical shortcoming in the communication abilities of many analytics professionals. Specifically, their ability to articulate in business terms what their analyses showed, and make actionable recommendations, usually left something to be desired. The common refrain we heard from middle and top management was that when analysts made presentations, they tended to lapse into the technicalities of mathematical procedures, rather than focusing on the strategic and tactical impact and meaning of their work. As analytics became more mainstream and widespread in organizations, this problem got more acute.

The genesis of this book was our desire to do something about this issue. We have experienced this on both sides, as presenters (not always successful!) and as members of an audience that got lost. Over the years, we experimented with different ways of presenting analytics work to make a more compelling case to top management. This resulted in our discovering some tried and true methods for improving one's presentations, a sampling of which we shared with different audiences, including corporate clients, academics, professionals, and conference attendees at the Institute for Operations Research and Management Science (INFORMS) Analytics Conference (2016), the Annual Conference of the Decision Sciences Institute (2017), among others. Audiences everywhere responded with enthusiasm and wanted more. Many suggested that we write a book about it. Additionally, they—both analysts and managers—wanted to share their own experiences too. This book is thus a collection of experiences—our own, and those of other academics and professionals involved with analytics.

The book is not a primer on how to draw the most beautiful charts and graphs or about how to perform any specific kind of analysis. Rather, it shares the experiences of professionals in various industries about how they present their analytics results effectively—in other words, how they tell their story to win over the audience. The reader will be able to use the ideas from the chapters immediately in their own presentations. The examples span multiple functional areas within a business, and in some cases, the authors discuss how they adapted the presentations to the needs of audiences at different levels of management.

We hope you enjoy the book, and we welcome any feedback.

Executive Summary

In Chapter 1, we discuss the importance of knowing your audience and the overall design of a presentation specifically suitable for top management. In Chapter 2, we demonstrate how the results from some common modeling techniques can be presented to a business audience.

Chapters 3–10 represent the contributions from analytics professionals who we invited to share their expertise. In Chapter 3, Jennifer Priestley, Associate Dean and Professor of Statistics and Data Science, shows how visualization can be used to improve the analytics process itself. Gregg Weldon, Co-founder and Chief Analytics Officer at Analytics IQ, a Marketing Data company, illustrates in Chapter 4 how the effectiveness of marketing analytics can be demonstrated to clients. Prof. William Swart, in Chapter 5, discusses how analytics can be presented to convince top management to bring about changes in operations in the fast-food restaurant industry. In Chapter 6, Col. Lynette Arnhart (U.S. Army, retired) discusses common types of analysis used in the armed forces and provides tips on presenting effectively to military leadership. Elyse Hallstrom, of Intel Corporation, discusses how analytics presentations can be successfully adapted for multiple layers of management with differing expectations in Chapter 7. Prof. Keith Miller, in Chapter 8, shows better ways to communicating process improvement techniques and results to executives. In Chapter 9, Jason Thogmartin, Senior Vice President

and Director of Audit at Santander US, presents actionable audit ana‑
lytics results for local and global management with different strategic
goals. J. P. James, a serial entrepreneur and board member at multiple
financial organizations, displays winning presentations to investors in
the consumer lending industry in Chapter 10.

Finally, in Chapter 11, we illustrate the principle of "less is more" in
making presentations. The focus is on keeping presentations unclut‑
tered and to the point. Every chapter in this book stands alone and
represents the experiences of various professionals regarding effective
presentations, with illustrations from their own industry. We believe
that readers will be able to use most of the ideas and examples
presented in this book in their work right away.

Editors

Subhashish Samaddar, PhD, Certified Analytics Professional (CAP®), is a professor of business analytics and operations management in the managerial sciences department of the J. Mack Robinson College of Business at Georgia State University, Atlanta, Georgia, where he was the founding academic director of MS analytics program. He served on the INFORMS' team that created the CAP certification examination administered globally and coedited its first guide book. An internationally reputed and multiple award-winning researcher, teacher, and speaker, he specializes in business analytics, operations and organizational knowledge management, and decision-making. A veteran of more than 25 years and a consultant in analytics, he has helped many U.S. organizations—Fortune 100, privately held and governmental agencies—with their analytical needs. He currently teaches business analytics and research methods to undergraduates, MBA and executive master's and doctoral students, and corporate clients.

 **Satish Nargundkar, PhD**, is a professor of business analytics in the J. Mack Robinson College of Business at Georgia State University, Atlanta, Georgia. Over the past three decades, he has helped large and small companies improve their decision-making through analytics. A recipient of multiple awards for teaching and research, he has over 25 years of experience in the areas of analytics, process improvement, and decision support. His research interests are multidisciplinary and include supply chain management, quantitative methods, and the improvement of teaching methods. He is passionate about excellence in teaching and is sought after as an instructor in executive programs. In his spare time, he enjoys reading, traveling, and photography, and is an instructor in martial arts.

Contributors

The contributors are listed in alphabetical order.

Col. Lynette M.B. Arnhart, PhD, U.S. Army, retired, is the former division chief for analysis, assessments, and requirements at United States Central Command and was responsible for developing and establishing the command's quarterly assessment of the Coalition Military Campaign Plan to defeat ISIS. She has significant experience at the headquarters, army, and combatant command level, and she has led analyses of all types—strategic assessments, human capital, weapon-system effectiveness, modeling and simulation, programming and budgeting, and decision analysis. Col. Arnhart has a PhD in operations research from George Mason University.

Elyse Hallstrom works as a data scientist at Intel. In her 11 years there, Hallstrom has moved from manufacturing quality control to substrate materials to decision engineering where she works today. She was named Technologist for her time in Materials based on her work in improving decision-making systems for the assembly test organization. She now works on improving decision-making processes and systems across the company. Hallstrom received her bachelor's degree in computer engineering from the University of Arizona, Tucson, Arizona and her master's degree in industrial engineering with a focus on operations research from the Arizona State University, Tempe, Arizona.

J.P. James currently serves as the chairman of the board at Libreum International LLC and is on the boards of several other companies. Over the past two decades, he has helped start, acquire, restructure, and scale stagnant ventures. He advised companies on their management strategy, business process automation, back office automation, and capital solutions. He built ventures from startup to hundreds of millions of dollars in valuation. He has invested in and advised fast-growing ventures since 2004. He has lectured on quantitative computational finance and eEntrepreneurship at Georgia Tech, Atlanta, Georgia.

Keith E. Miller, PhD, has two decades of experience as a military officer, management consultant, corporate master trainer, and Six Sigma practitioner and has led or coached more than 400 process improvement projects to completion. He is a certified Six Sigma Black Belt and Master Black Belt, and a project management professional. He has an MS and PhD in chemical engineering from Northwestern University, Evanston, Illinois, and currently teaches at Clayton State University, Morrow, Georgia. His research interests include process improvement methodologies, business analytics, and interactive business pedagogy.

Jennifer Lewis Priestley, PhD, is the associate dean of The Graduate College and the director of the Analytics and Data Science Institute at the Kennesaw State University, Kennesaw, Georgia. Prior to receiving a PhD in decision sciences from Georgia State University, Atlanta, Georgia. Dr. Priestley worked in the financial services industry for 11 years. Her positions included vice president of business development for VISA EU in London, as well as for MasterCard US. She also worked as an analytical consultant with Accenture's strategic services group. She holds an MBA from Penn State, State College, Pennsylvania and a BS from Georgia Tech, Atlanta, Georgia.

William Swart is a professor of marketing and supply chain management at the East Carolina University, Greenville, North Carolina. He has served as the vice president for operations systems at Burger King Corporation as well as Provost at East Carolina University. He has twice been a finalist for the INFORMS-sponsored Franz Edelman competition for the best application of OR/MS in the world. He has received the IIE Operations Research Division Practice Award. He has also received the NASA/KSC Group Achievement Award for his contributions to space shuttle ground processing. He received a PhD in operations research from the Georgia Institute of Technology, Atlanta, Georgia.

Jason Thogmartin, DBA, is the senior vice president and director of audit at Santander, a United States' leading audit coverage of the retail banking business and financial, operational, and compliance risk areas. Prior to joining Santander, Thogmartin held roles at First Data as the head of audit for the firm's global lines of business and shared services functions. Prior to joining First Data, Thogmartin held several roles with GE, GE Capital, and NBC Universal. Within GE Capital, he was managing director and head of audit for corporate and previously managing director and head of audit for retail and commercial banking in the Americas.

Gregg Weldon is currently the co-founder and chief analytical officer at AnalyticsIQ, a marketing data company that specializes in creating individual, household, and area-level marketing and demographic data that allow its clients to identify potential customers via both traditional and digital platforms. Prior to that, he was a vice president of modeling at Sigma Analytics, helping with the transition of the company's focus from risk models to marketing models. He began his career building risk models for financial organizations at CCN Inc. and Equifax. He has an MA in economics from the University of South Carolina, Columbia, South Carolina.

1

KNOW YOUR AUDIENCE

SUBHASHISH SAMADDAR AND SATISH NARGUNDKAR

Georgia State University

Contents

I (Subhashish Samaddar) was scheduled to make an hour-long presentation to the senior management of a client company regarding the optimization of retirement benefits to be paid out to their employees. This was a project I had been working on for about a year with this client. My recommendation based on detailed financial impact analysis had to do with whether to maintain the status quo or to implement a new multilayered postretirement benefits package. Immediately prior to the meeting, I had lunch with several members of senior management from the client company, including the Senior Vice President (SVP) of Human Resources, who was ultimately responsible for the decision. As the lunch was coming to a close, and I was about to start my presentation, an employee walked up to the SVP and whispered something in his ear. There was a pause in the room, and I could sense from the SVP's body language that something was wrong. Confirming my suspicion, he walked over to me and said,

> I need to take care of an emergency situation and leave within a few minutes. Would you like to reschedule your presentation for another day or give me the highlights of your analysis and recommendation right now? I can give you about five minutes.

Not wanting to leave things hanging and uncertain, I opted to speak to him right then. I had just a few seconds to distill the entire presentation down to my key recommendation and the most compelling arguments to support it. I got it done in four minutes. He smiled, looked at his second-in-command, and said,

"I like it. I am in. Why don't you continue and let me know what you think." He then left the room in a hurry.

The rest of the meeting lasted only about 15 minutes. All of my recommendations were accepted for implementation.

On the way back home that day, I realized that I got lucky. I was able in this instance to do what I did on the fly, but that may not always work. I made a note to myself that in the future, I would always be prepared with a 3–5-minute summary of any presentation that I planned to give, and make sure it was compelling enough to seal the deal. I have learned that shortening the presentation down to its most essential elements takes significant thought and preparation. The good news is that we often know what is necessary to sell our ideas. The bad news is that we know a whole lot more about the problem and its analysis and are eager to share all of that wealth of knowledge with any audience, assuming that they *need* it. The question to ask oneself is, "Is it really necessary to say all this to this particular audience to achieve my objective?" Over the years, I have developed a rule to help me be more succinct. I simply ask myself the question, "Would I still include this sentence in the presentation if it cost me $1,000 to do so?"

Most well-known speakers in history have in some form expressed the idea that it takes more effort to make a speech short and effective. A famous quote, variously attributed to Woodrow Wilson, Winston Churchill, and others, goes something like this—"If I am to speak ten minutes, I need a week for preparation; if an hour, I am ready now." Make sure you spend that "week" to prepare.

Preparing for Your Presentation

During an analytics project, you (the analyst) generally will make more than one presentation, during various phases of the project, to different stakeholders. To prepare for an effective presentation to senior management, the first thing one needs to do is to define the objective clearly. What are you trying to achieve? How will you know if your presentation was a successful one? The more specific you can make the objective in your own mind, the more likely you are to achieve it. For most analytics professionals, the objective of a presentation tends to be to convince management that the analysis meaningfully addresses a business problem, and to make recommendations for action, along with compelling evidence that makes it likely to get their buy-in.

Once the objective is clear, the next step is to understand the audience. Who will be present? What are their objectives? What are their expectations coming in to this meeting? What is their professional or academic background? Members of senior management are generally highly intelligent and capable. They are well informed about their business and are generally well educated, though not necessarily in analytics or related quantitative fields. Be careful not to underestimate someone because they do not have your technical background. They are generally busy and impatient with details. They have a wider view of the organization and are engaged in making many high-stakes decisions. The problem addressed by your analysis is likely to be just one of them. The challenge for you is to present your technical analysis and recommendations in a way that is accessible and relevant to your audience.

Source: http://rankmaniac2012-caltech.blogspot.com/

It is worthwhile to use all the channels available to you to find out what you can about your audience prior to designing the presentation. Aside from their objectives and who they are, it can help to find out their preference in processing information. For instance, some may be more comfortable with charts and figures, while others may prefer a presentation without any visual aids (some managers have a "no PowerPoint" rule). It helps to discover any special capabilities, preferences, or needs of your audience. For instance, some managers that are technically savvy may want to explore the data/models you present to test assumptions or explore scenarios. In such cases, it might be necessary to be prepared to have a demonstration ready on some software that they can use to work along with you. An example of a special need could be audience members that are color blind. This will impact your choice of visuals. Knowing the risk tolerance of the audience can also affect the recommendations that you present. Entrepreneurs and angel investors may be more risk tolerant than top management of an established organization.

Organizing the Presentation

The first task in organizing your presentation is to decide the order in which to tell the story. The typical approach used by many analysts is the chronological sequence, beginning with an introduction, data and assumptions, the analytical methodology used, the results, and to finish, the recommendations and conclusion (Table 1.1).

This sequence is ingrained in our minds, perhaps for a couple of reasons. One, most college project or research presentations and academic papers follow this structure. Also, this is the sequence in which the work is done, so it seems logical to present it that way. An academic audience may appreciate this sequencing, since in the academic world, it is critical to show the development of the thought process behind the work done and to demonstrate that the methodology used is appropriate. The result itself is not paramount. The opposite is often true in the business world. The business client is primarily interested in results,

Table 1.1   Traditional Presentation Sequence

PRESENTATION ORDER 1
Introduction
Assumptions
Analysis
Results
Recommendations
Conclusion

Table 1.2 More Effective Presentation Sequence

PRESENTATION ORDER 1	PRESENTATION ORDER 2 (MORE EFFECTIVE)
Introduction	Introduction
Assumptions	Recommendations
Analysis	Results
Results	Analysis
Recommendations	Assumptions
Conclusion	Conclusion

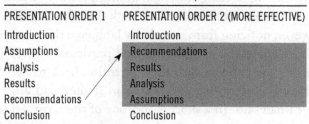

not in how your mind works or the process you went through to get there. This is not to imply that using the appropriate methodology is unimportant in business. The point is that if you have been entrusted with the task and are assumed to be competent at it, therefore, you know the right methods to use. The methodological details do not need to be presented unless asked for. While buying a custom computer desk, we are interested in the features and best use of the finished product, not in what tools and processes the carpenter used to build it.

In our experience, a more effective sequence for presentations to senior management is to get straight to the business recommendations after the introduction, justify them with the results of your analysis, and discuss the methodology only if asked (Table 1.2). However, always be prepared for that contingency. It is useful to have a separate presentation handy in case someone is interested in the complexities of the techniques, algorithms, data, or process used.

You can experiment with variations of this theme depending on the context and your audience. For instance, you may find it useful to discuss the assumptions first, and then go to the recommendations. Alternately, would it be more beneficial to show the results before making recommendations? While we found the sequence shown above to be generally the most effective, the key is to be flexible enough to design the presentation based on your objectives and your understanding of your audience.

Audience Interaction

The discussion above deals with what you do before the presentation. Now let's discuss the presentation itself. A while ago, about 15 minutes into an hour-long presentation to a client, I noticed that some of the senior audience members had stopped listening to me. They were animatedly talking to each other and pointing to the screen, ignoring my presence. My first reaction was to think that I had made some mistake. However, noticing from their body language that they seemed to have obtained some key insight from that particular slide, I remained silent, waiting for them to turn their attention back to me. I resisted the temptation to intervene and ask them if they had any questions about what I had said. In a short while, one of them turned to me and said with a smile "now we know the answer to our problem." I was a bit surprised, and before I could say anything, he said, "Unless you have something else to present, we are done. Can we break for lunch now?" I was happy to end the presentation right there and go to lunch with them 40 minutes early! The audience's broader knowledge of their organization complemented my presentation and helped them get an insight from my chart that I could not have delivered. The lesson I learned was that one cannot always fully anticipate how the audience will react to something. Paying attention to the audience's body language and remaining flexible was the key. However, that is difficult to do without first being at ease with oneself and one's presentation.

In the situation mentioned above, I briefly lost the audience for what turned out to be a positive reason. That is of course not always the case. In a different presentation, I noticed negative body language from a senior analyst in the audience, who I knew had a master's degree in statistics. Recognizing that he may have had a concern that he was not voicing, I stopped and asked him if he had a question or comment. It turned out that he did have a technical concern and strongly disagreed with me. I was able to resolve the situation to his satisfaction and win him over to my way of thinking by going into some technical details that I would otherwise not have presented. It was important, however, that I was prepared for such a contingency. Needless to say, a complete grasp of the technical aspects of what you are presenting goes a long way in making an effective presentation.

It is good practice to spend some time thinking through your presentation beforehand and anticipating possible questions or challenges. When possible, preempt potential questions by building the answers into your presentation. Additionally, be cognizant of the human side to presentations. Humility, a sense of humor, and simply being comfortable in conversations with people can be a big asset in achieving your objectives in any presentation.

2

PRESENTING RESULTS FROM COMMONLY USED MODELING TECHNIQUES

SUBHASHISH SAMADDAR AND SATISH NARGUNDKAR

Georgia State University

Contents

As an analytics professional, one may use a wide variety of mathematical techniques. Each of these techniques has its own mathematical algorithm, and the output is often filled with obscure details that are valuable to the analyst, but a distraction to a managerial audience. Distilling the key insights from the output and presenting them in the language of business rather than that of mathematics is crucial. In this chapter, we illustrate this idea by suggesting some ways to present the results from two commonly used techniques—regression analysis (for prediction) and cluster analysis (for segmentation). The numbers and variables used in these examples resemble real-life business cases, but are fictitious.

Regression Analysis

Some variant of regression analysis is generally used for predictive modeling. Multiple regression, linear discriminant analysis, logistic regression, and other techniques are at their core quite similar to each other. We discuss here an example of output from multiple regression that was used to predict the Sales ($ thousand) based on Advertising

Expenses ($ thousand), Bonus ($) paid to salespeople, and Sales Force (number of salespeople). This output is from Microsoft Excel®. Consider that you are presenting the results from this analysis to your client.

Many analysts simply take the output as is, display it on the screen (Table 2.1), and talk through the details, maybe with the help of a laser pointer.

This kind of table crowded with numbers is an immediate turn off for a nontechnical audience. Further, showing all the numbers at once distracts the audience from what is most important about the output, which is the model itself. One reason presenters do this is to have all the information (R-square, p-values, F-statistic, ANOVA table, standard error, etc.) in one place, and the presenters are tempted to discuss every element from one display. Most of the time, the audience may not be interested in that information. Even if they need to know some of this information, it generally requires further interpretation to make it accessible. This can create an information overload. An immediate improvement in presentation is to first present only the model as shown in Table 2.2 instead.

Table 2.1 Regression Output

SUMMARY OUTPUT

REGRESSION STATISTICS

Multiple R	0.99726				
R-square	0.994534				
Adjusted R-square	0.99271				
Standard error	16.97083				
Observations	13				

ANOVA

	DF	SS	MS	F	SIGNIFICANCE F
Regression	3	471638.72	157212.9	545.8608	1.7E-10
Residual	9	2592.08	288.0091		
Total	12	474230.80			

	COEFFICIENTS	STANDARD ERROR	T STAT	P-VALUE	
Intercept	−119.706	53.41416	−2.24109	0.051751	
Ad Expense	22.38734	9.269338	2.415203	0.038916	
Bonus	0.572689	0.228156	2.510083	0.033306	
Sales Force	66.85939	12.28413	5.442745	0.000410	

Table 2.2 The Prediction Model

Predicted sales ($000) = −119.706 + 22.38734 (AdExp $000) + 0.57269 (Bonus $) + 66.85939 (SalesForce)

This gets to the essence of the matter, which is to show the variables that significantly affect the predicted variable, and the mathematical relationship between them. However, some improvement is still possible. It is still too mathematical for some and needs a clearer business interpretation. The predictor variables are measured in different units, and the coefficients must be appropriately interpreted to make sense of what the model is suggesting. Tables 2.3 and 2.4 show an alternate way to present the results.

This looks similar to a section of the original output in Table 2.1, but explains the meaning of the coefficients in the column heading, rather than simply calling them coefficients. It may be necessary to explain that the effects shown in the model assume that you are changing one variable at a time, holding all else constant. Note, however, that Table 2.3 does not write out the complete equation as in Table 2.2. One might argue that some information is lost in doing so. Without the intercept and the complete equation, one cannot actually find the predicted value. However, remember that the actual prediction will be done in some automated system when the model is implemented. What management most likely needs to know during

Table 2.3 Relationships Interpreted

VARIABLE	CONTRIBUTION TO SALES ($000) PER UNIT INCREASE IN THE PREDICTOR
Ad Expense ($000)	22.40
Bonus ($)	0.57
Sales Force (number of people)	66.86

Table 2.4 Interpretation in Sales Dollars

VARIABLE	CONTRIBUTION TO SALES ($) PER UNIT INCREASE IN THE PREDICTOR
Ad Expense ($000)	$22,400
Bonus ($)	$570
Sales Force (number of people)	$66,860

the presentation are the key relationships, and not the exact equation to be implemented. It may be preferable for some to show the actual dollar amounts of Sales instead of in thousands of dollars. That is an easy switch, as shown in Table 2.4.

One could stop here as far as presenting the model itself. However, it is likely that you have been hired to do this analysis to help management decide where to invest money in order to increase sales. If so, the relationships as presented do not really permit an easy comparison of the alternatives. This is because the units used to measure the predictor variables are all different—thousands of dollars for Ad Expense, dollars for Bonus amount, and a count for the Sales Force. Direct comparison can be facilitated by equalizing the units for the variables, preferably in dollar amounts. In other words, if one had an extra dollar to invest in one of the three areas, what would give the most bang for the buck? A casual inspection of Tables 2.3 and 2.4 might mislead some into believing that Sales Force has the biggest impact ($66,860). Table 2.5 makes the comparison clearer.

At this point, someone in the audience will likely ask (and this is a good sign, implying they have been following you!) you how good these numbers are. How low or high can they go? It helps to mention here that the numbers shown in Table 2.5 represent the expected influence on average. You can at this point be prepared with a backup slide that shows the confidence intervals (typically 95%) of the estimates, presented as best and worst case scenarios (Table 2.6).

While presenting this, one could add that there is a 95% chance that the effect will lie within the limits shown. In addition, one might ask how far the model itself will hold true. In other words, to what extent can one keep investing in these three resources and still expect the same levels of effect on Sales? This has to do with the range of

Table 2.5 Increase in Sales ($) per Dollar Increase in Resource

RESOURCES	CONTRIBUTION TO SALES ($) PER $1 INCREASE IN RESOURCES
Ad Expense	$22.40
Bonus	$570.00
Sales Force (number of people)	$0.66[a]

$1 spent on bonus has the highest return ($570), all else held constant.
[a] Assumes a full-time sales person costs $100,000.

Table 2.6   What Are the Limits of the Estimated Contributions?

RESOURCES	WORST CASE	CONTRIBUTION TO SALES ($) PER $1 INCREASE IN RESOURCES	BEST CASE
Ad Expense	20.00	$22.40	24.80
Bonus	540.00	$570.00	600.00
Sales Force (number of people)	0.59	$0.66	0.73

values for the predictors used in the data. We do not show it here, but it is useful to have the range of possible values for each predictor variable handy, since a regression model only works reliably within that range.

Finally, it is possible that the client believes, based on their past experience or gut feel, that Salespersons Salaries should have been in the model and asks you why it is not. It is important to keep track of what variables were dropped from the model and why. A seemingly useful variable often drops out due to multicollinearity, something that can be explained as shown in Figure 2.1.

Assume that the large circle shows the total variation in Sales that you are attempting to explain with the model. Each predictor variable is correlated with Sales to a different degree, shown by the overlap of its circle with the Sales circle. Salesperson Salaries has a considerable overlap with Sales, but the portion of Sales that is explained by the Salary variable is already explained by the Sales Force and Bonus variables. In other words, there is a strong correlation between Salaries,

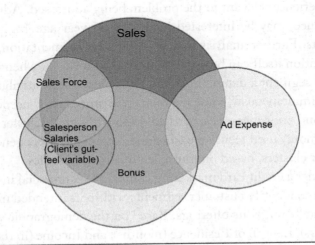

Figure 2.1   Multicollinearity.

Sales Force, and Bonus. Salaries therefore adds no useful information in explaining or predicting Sales, beyond what is already known based on the variables in the model already presented. A follow-up question from the audience might be why one of the other two variables, that is, Sales Force or Bonus, was not dropped from the model instead. As shown in the figure, both of those variables have a larger overlap with Sales, and thus, dropping one of them instead of Salaries would reduce the overall proportion of Sales explained by the model. Analysts may note that this argument is essentially addressing the impact of the variables on R-square, without using that term.

Cluster Analysis

When data need to be segmented, one way is to simply use categorical variables that automatically divide the data into categories or segments. For instance, Sales data may be segmented on the basis of salesperson, or product type, or region. Demographic data may be segmented by gender or ethnicity. Numeric variables can also be used to segment data. A dataset can be broken down by age, for instance, into two segments: those over 35 and those at or under 35. This is easy to do mentally with a single numeric variable, or perhaps two variables considered together, such as Age and Income. In other words, no mathematical analysis is needed to make the decision about where to split (in this case, age 35). It can be done on the basis of one's business experience relevant to the problem being addressed. Advertisers, for instance, may be interested in the 18–35-year age group's viewing habits. Further analysis may be done after segmentation, but the segmentation itself can be done subjectively. However, when it is necessary to segment a dataset based on multiple numeric attributes considered simultaneously, some mathematical procedure is needed, since it is beyond one's mental capacity to do so without it. Cluster analysis is a commonly used technique to segment data into homogenous subgroups or clusters, based on multiple numerical variables.

Consider a credit card marketing campaign. A financial institution hires you to identify customer segments within its intended market to appropriately target its offerings. Based on the demographic variables Age (years), Length of Residence (months), and Income (in thousands of dollars), and the credit variables Number of Credit Accounts and

Credit Utilization (%), you run a *K*-Means clustering procedure. This approach requires going through a process that involves experimenting with various number of clusters and using some statistical guidelines, rules of thumb, and interpretability to converge on a solution with a certain number (*K*) of clusters.

When presenting the results to the client, it is best not to speak of the entire process unless specifically asked to, and hone in instead on the final solution. In this case, assume Table 2.7 shows the final solution.

For the reader not familiar with cluster analysis, each column in the table shows the average values of the variables for all the customers within the corresponding cluster, called the cluster centroid. For example, the average age of the customers in Cluster A is 35, with an average Length of Residence of 72 months, and an average income of $50,000 per year. They have on average 18 accounts each, with a credit utilization of 73%. However, explaining each cluster by going over each of the 25 numbers shown in the table is cumbersome and is likely to distract the audience. It also risks missing out on the bigger picture that the clusters paint. It is better to interpret each cluster in broad terms. For instance, compared to the overall sample, people in Cluster A are relatively young and are risky prospects, while those in Cluster B are possibly retirees with no interest in more credit. Table 2.8 shows one way to present the interpretation of each of the clusters. It may even be useful to present this table to the client first and then show the details of the centroid values from Table 2.7 as supporting information.

Table 2.8 should lead you to the crux of the presentation, which will be to highlight the business implications that your client wants to know about. What are your recommendations about the targeting of customers for the credit card marketing campaign? Answering this

Table 2.7 Final Cluster Solution

	CLUSTER A	CLUSTER B	CLUSTER C	CLUSTER D	CLUSTER E	TOTAL
Age	35	65	50	45	55	54
Residence	72	240	180	144	160	178
Income	50	45	95	75	80	62
Accounts	18	3	15	8	5	8
Utilization	73	5	25	45	15	27

Table 2.8 Cluster Interpretations

CLUSTER	CUSTOMER CHARACTERISTICS
Cluster A	Young, risky
Cluster B	Retired, not interested in credit
Cluster C	Wealthy, moderated users of credit
Cluster D	Prosperous, good prospects for new credit
Cluster E	Prosperous but unprofitable users

typically requires further analysis within the segments. Predictive models can be built for each segment to come up with scores or probabilities of response or customer risk. Table 2.9 shows the results of such analyses for this example.

As the table shows, people in Cluster A, roughly 15% of the population, are the most likely to respond, but show negative profitability, and are thus not a good target. If the client chooses to target them anyway, this analysis can at least provide them with an understanding of the risk involved and permit them to mitigate that risk with suitable strategies. People in Cluster B are the least risky, but are unlikely to respond at all, and again, are not a good target. This helps the client eliminate 40% of the sample from consideration for the marketing campaign, significantly reducing cost. People in Clusters C and D, a total of about 30% of the population, seem to show the best balance between likelihood of response and potential for profit. Thus, targeting those people makes for a much better targeted marketing campaign. It creates significant savings in dollars spent on the campaign, while getting most of the desired customers from the population, thus maximizing the effectiveness of the campaign.

The information in Table 2.9 can also be illustrated in graphical form as shown in Figure 2.2. The five clusters are placed along two dimensions—likelihood of response and profitability. The size of the

Table 2.9 Predictive Models within Clusters

CLUSTER	VOLUME (POP. %)	RESPONSE PROBABILITY	AVERAGE PROFIT
A—Credit dependent	15	0.35	(75)
B—Shuffle board set	40	0.01	5
C—Country club set	10	0.08	48
D—Prosperous high profit	20	0.13	86
E—Prosperous low profit	15	0.03	(5)

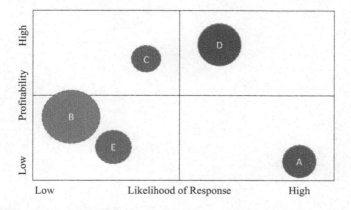

Figure 2.2 Cluster evaluation for marketing campaign.

circle represents the size of the cluster. This way of presenting it also gets straight to the heart of the matter—key characteristics of the various segments of the customer base—allowing the decision-maker to focus on the business of marketing, instead of on the statistical minutiae.

Summary

The two examples above, using regression and cluster analyses, are intended to demonstrate ways to present the results of your analyses to middle or upper management. The essence of a successful presentation is to ensure that you give your audience the main message first—what action can they take? That is, what does all the analysis ultimately mean to the client in terms of making business decisions? In both the examples presented here, the final recommendation to the client was about where they should spend their money or focus their efforts to most effectively achieve their business objectives. One can drill down into the details of the analysis as needed for each audience, depending on their technical expertise and interest. However, starting with such details and building up to the business implication is usually the wrong way to go.

3

VISUALIZATION TO IMPROVE ANALYTICS

JENNIFER LEWIS PRIESTLEY

Kennesaw State University

Contents

The Paradox of Visualization

In data analytics, visualizations are consistently underutilized. While visualizations like bar charts and bubble plots are now well-worn tools within most communication toolboxes, these tools are typically used to communicate concepts such as trends, magnitude, or relativity associated with structured results at the conclusion of an analytical process—sort of the cherry on the top of the project. Consider the outline of typical tasks associated with data science depicted in Figure 3.1.

Visualization is what happens at the end of an analytical process—typically the last-minute task that one executes before a presentation to the boss. Too often the individuals engaged in the more computationally

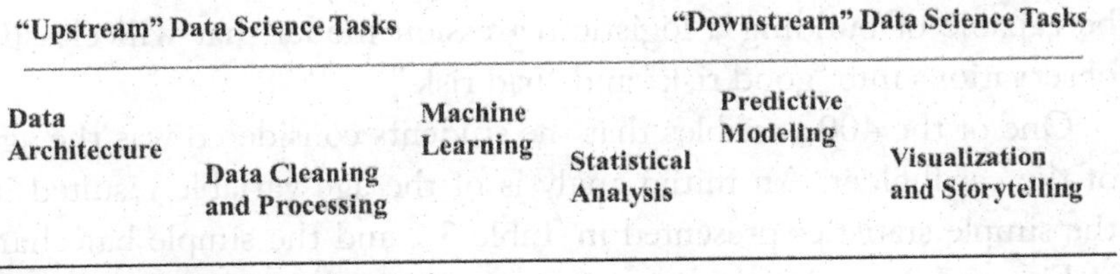

"Upstream" Data Science Tasks			"Downstream" Data Science Tasks	
Data Architecture		Machine Learning	Predictive Modeling	
	Data Cleaning and Processing		Statistical Analysis	Visualization and Storytelling

Figure 3.1 Typical data science tasks.

17

rigorous tasks—in Figure 3.1, the "upstream" tasks—see the visualization (and associated storytelling) of analytical results as not their responsibility. They often consider it the most "simplistic" part of the process and of little value to their work. This dismissal of the value that visualizations can bring to a project as merely the simplistic aspects of the analytical process contributes to data scientists and analysts possibly creating suboptimal models and potentially generating lower profits for their organizations. This is the *Paradox of Visualizations*—those individuals who are the most computationally skilled (e.g., computer scientists and data scientists) are the most likely to dismiss the value of visualization tools in the analytical process as too simple and unnecessary. The corollary to this paradox is that frequently, the less computationally skilled individuals are the most likely to appreciate the value of visualizations but least likely to have the knowledge to understand the complexity well enough to simplify it. In other words, what is needed is analysts competent enough to understand all of the complexity, yet mindful enough to express the essence of the analysis in simple terms.

In this chapter, I highlight the underappreciated value of using visualizations of data as a tool to inform and improve the processes related to actually identifying which data should be used in a process, what form that data should take, and how to refine and iteratively improve model performance.

Things Are Not Always as They Seem

Consider a typical credit risk-modeling project. A subprime lending institution approached our university's analytical research center, looking to improve the accuracy of their risk models, and ultimately increase their profitability. They provided us with about 18 million accounts and about 400 credit attributes. Using these data, they asked us to develop a series of risk models. The request seems simple enough—any graduate student in a computational discipline should be capable of building a logistic regression model that will classify observations into "good risk" and "bad risk."

One of the 400 variables that the students considered was the age of the cardholder. An initial analysis of the age variable resulted in the simple statistics presented in Table 3.1 and the simple bar chart in Figure 3.2.

Table 3.1 Descriptive Statistics for Cardholder Age by Good Risk and Bad Risk

RISK	N	MINIMUM	MEDIAN	MEAN	MAXIMUM
Good	14,357,982	21	48	56	99
Bad	3,927,361	21	47	62	99

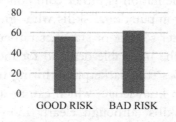

Figure 3.2 Average cardholder age by credit risk classification.

The students executed similar tables and graphs for several of the potential predictive variables. From their perspective, the creation of these graphs was necessary only to communicate the descriptive statistics, rather than to diagnose potential issues with the data that could confound the modeling process. Since they thought of the visuals as a simplification for delivery to someone else (the client) rather than for them to use as diagnostic tools to inform and improve their modeling, they missed identifying key problems in the data for that variable. Similarly, they had 400 such variables across the dataset with potential problems that they could miss.

The Role of Domain Knowledge

There are two issues with what they presented. The first is that the reported means of the age variable—56 and 62—seem high, certainly relative to the median values of 48 and 47. Given that the average age in the US adult population is just under 40, these values appear to be illogical (although 56 and 62 seem increasingly young to me...but I digress). The second issue is that the maximum value is 99 for both groups. While that number could be true, it seems contrived. Why not 98? Why not 101? Why equal for both groups?

Although these students likely had credit cards of their own, it is important to note that they had limited *domain knowledge*—none of these students had worked in consumer finance or in risk. The analytics program where these students are enrolled emphasizes mathematics, machine learning, statistics, and programming, but places less emphasis on domain knowledge. Put simply, the data are just "xs and ys" to them. It would be difficult without any domain knowledge to identify that there are potential problems with the values in Table 3.1.

This "x and y" orientation is rather common for individuals with particularly strong computational skills who "grew up" (analytically speaking) in computer science or statistics rather than in finance or health care. Given the insatiable demand for individuals with deep analytical skills across the marketplace, hiring managers frequently will sacrifice domain knowledge, which can be acquired on the job, for computational skills (although clearly individuals with strong computational skills and domain knowledge will always be preferred). An important caution for managers in this position is to be aware that individuals with strong computational skills but limited domain knowledge are more likely to miss "obvious" issues within the data and/or "discover" trivial patterns. For instance, an international student, unaware of Halloween in the United States, brought to my attention his "discovery" that a candy retailer experienced a spike in sales in late October.

I asked them to go back and create alternative visualizations of age—this time for *them* rather than for the client. There was something going on with the data in Table 3.1 that was not readily apparent through a simple investigation of the descriptive statistics. More precisely, I reminded them of the lowly histogram that they learned as undergraduates (to which they rolled their eyes as only graduate students can). The histogram of *Age* in Figure 3.3 provided the students with a little more insight into the variable.

From this histogram, the issue becomes clearer—the values of 99 on the far right of the graphic are out of logical expectation relative to the rest of the distribution of the variable. While it is *possible* that the data include a group of centenarians applying for subprime credit cards, that scenario is not likely. Further investigation of the other 399 variables indicated similar spikes at the 99 value. The students did not need vast amounts of domain knowledge to see that these results were, at best, statistically unlikely and should be investigated, but at worst they were just wrong—their inclusion could substantially decrease the accuracy and validity of their models. However, what the

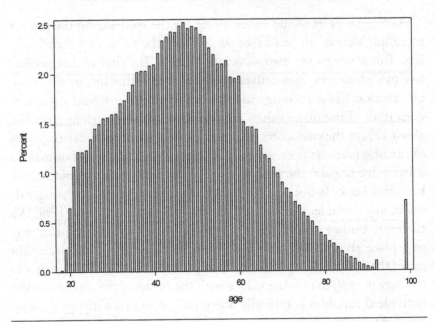

Figure 3.3   Histogram for the cardholder age.

students needed was a reminder that visualization is a tool that they can and should use to inform and improve their understanding of the behavior of the data.

Moving through Complexity

While a histogram is in itself not "complex," the use of a histogram for the purposes of evaluating and diagnosing potential issues with data is a more complex utilization of the visualization as a tool for diagnosis, rather than as a "nice to have." This is particularly important when the analyst has limited domain knowledge.

After a conversation with the client, we determined that the values of "99" were actually codes to indicate that the data were missing rather than indicating an age of 99 (this is common practice, particularly in consumer credit applications). Importantly, because these missing values were coded as numbers, but were not part of the scale of the variable, they were wrongly included in the calculations of the descriptive statistics in Table 3.1 and would have been almost impossible to recognize, even with domain knowledge, without visual inspection. This was true across all 400 variables.

At this point, it would be tempting for the students on the project to simply delete any values of 99 and set the values to *truly* missing. This strategy has two associated issues. The first is that predictive modeling processes utilize complete cases. This means that if an observation has a missing value for a variable, we would discard it even if all of the other values are populated. In the current example, about 17% of the customers have no reported age. While deleting 17% of the observations from a total dataset of 18 million may not make a substantive impact, the problem is that all of the variables potentially have this issue. If the analyst deletes all customers with missing values in any variable, the loss of data would be far more than 17%. We therefore cannot simply delete the observations. We must find a way to replace the missing values with something logical. Replacement raises the second issue related to these missing values. While a simple strategy to replace missing values with the mean or the median of the individual variables is typically acceptable, it may not always address the issue appropriately.

I asked the students to create distributions of the missing values by risk classification and by age. The simple visual in Figure 3.4 was very informative.

As shown in Figure 3.4, customers classified as good credit risks were substantively less likely to have values for age missing relative to bad credit risks. This fact was not apparent from the simple descriptive statistics in Table 3.1. The visualization helped to inform the students regarding an appropriate strategy for the replacement of the missing values (they used a stratified median-based imputation, but they also created unique categories for age = missing). Table 3.2 shows the revised descriptive statistics, and Figure 3.5 shows the revised

Figure 3.4   Percent of age values missing by credit risk.

Table 3.2 Descriptive Statistics for Cardholder Age with Missing Values Imputed

RISK	N	IMPUTED VALUES	MINIMUM	MEDIAN	MEAN	MAXIMUM
Good	14,357,982	2,153,697	21	48	48	92
Bad	3,927,361	981,840	21	47	50	91
Total	18,285,343	3,135,538	21	48	48	92

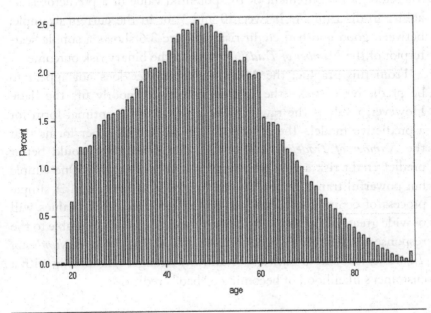

Figure 3.5 Revised histogram for the age variable.

histogram with the age variable completely populated—missing values replaced with values derived from information related to the age category and the risk classification.

Could these students have identified these issues without the aid of the visualizations? With substantive domain expertise, they would have had an expectation regarding the behavior of the data. However, with little or no domain expertise, regardless of the depth of their computational skills, the likelihood of identifying these issues was very low.

A second example of the relevance of visualizations to those engaged in the "upstream" tasks of analytics involves a second important variable in the credit scoring process—*Number of Tradelines* (the number of credit accounts a person has—each loan, credit card, etc. counts as a tradeline in one's credit history).

In logistic regression, the eventual predictive model will take the form of a sigmoidal function:

$$F(x) = \frac{e^x}{1 + e^x}$$

Therefore, a measurement of the potential value of a predictor is its ability to discriminate between 0 and 1, or in the current example, between good and bad credit risks. Figure 3.6 shows a simple scatterplot of the *Number of Tradelines* against the binary risk outcome.

From this graphic, the *Number of Tradelines* does not appear to be predictive of risk—the sigmoidal function poorly fits the data. However, rarely is the raw form of a variable the optimal form for a predictive model. The students then considered other forms that the *Number of Tradelines* could take that potentially would better predict credit risk and then assess these forms visually. One simple but powerful transformation is creating ordinal categories. A simple process of converting a continuous variable into ordinal values will provide greater understanding of the relationship of the variable to the response variable. In Figure 3.7 below, the ordinal form of *Number of Tradelines* is found to have a substantively stronger relationship with a customer's likelihood of becoming a "bad" credit risk.

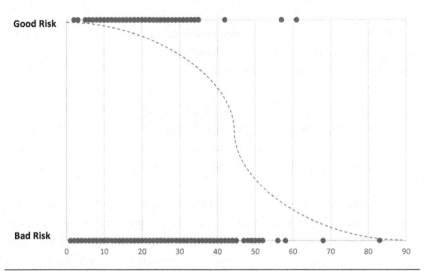

Figure 3.6 Relationship of *Number of Tradelines* to credit risk.

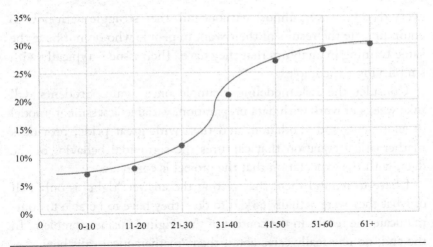

Figure 3.7   Percent of bad credit risks by ordinal categories of tradelines.

The students went through this process for all 400 potential predictors. As they developed some facility with the data—as they were building some level of domain expertise in real time—they were able to anticipate expected distributions, percentage of missing values, optimal replacement strategies and optimal transformations. After working with a dozen or so variables "manually," they were able to develop a "cleansing macro" to automate (simplify) the process. However—and this is important—they had to earn the simplicity of an automated macro. Without taking the time to examine the visualized distributions of the data, identify missing values, assess the optimal method for replacing these missing values, and evaluate transformation options, their macros would have likely been impressive, but wrong. Students and early career analysts are quick to want to take shortcuts to "macrotize" processes—particularly those students and analysts with highly computational degrees. Slowing down, leaning into the visualizations, and learning the data are some things that project managers need to remind them to do.

Simplicity Is Hard

As the director of a data science program, I have the privilege of working with highly computational students who can do impressive things with data—I learn new applications and techniques from them almost every day. However, almost without fail, they struggle with how to communicate the results of their work to people who do not have the same training in analytics that they have. Their issue is typically with simplifying the complex.

Consider the risk-modeling example once again. Students will take weeks of work with data preparation, variable assessment, model development, and validation, and then with great pride, provide a mathematical equation that captures the sigmoidal behavior of the data, with the expectation that the project is complete.

I have to remind them that getting the answer "right" is only half of what they were actually tasked to do—they have to be able to communicate the results in the context of the original business problem. In the end, no one really cares about their "mathematical calisthenics"—people want to understand how their results can benefit them and their company. Again, this is more than just creating a bubble plot. This last phase of any analytical project—the far right side of the tasks from Figure 3.1—is sometimes the most challenging for the early-stage analyst or for the highly computational student.

Consider the function that I received from the graduate students who engaged in the credit risk-modeling project:

$$\text{logit}(Y) = \beta_0 + \beta_1 \left(\text{Age of oldest line}\right)$$

$$+ \beta_2 \left(\text{Highest revolving balance}\right)$$

$$+ \beta_3 \left(\text{Income to debt ratio}\right)\ldots$$

The associated parameter estimates and significance are provided in Table 3.3.

While this detailed level of output is necessary for the eventual operationalization of the model, this is clearly not what needs to be communicated to the client. Specifically, individuals with strong computational backgrounds get very excited about beta coefficients and p-values; however, explanations of results which include these terms may as well be written in Martian for the average client. However, simply communicating complex concepts is difficult. Typically, I encourage my students to start simple and build up to the complexity rather than the other way around.

Table 3.3 Maximum-Likelihood Parameter Estimates from Logistic Regression

PARAMETER	DF	ESTIMATE	STANDARD ERROR	WALD CHI-SQUARE	PR > CHISQ	STANDARDIZED BETA ESTIMATES
Intercept	1	−0.8353	0.0168	2,464.6617	<0.0001	
Age oldest line	1	0.1330	0.00179	5,531.9121	<0.0001	0.1718
Highest rev balance	1	0.1138	0.00205	3,095.6583	<0.0001	0.1501
Income to debt ratio	1	0.0458	0.00101	2,053.9485	<0.0001	0.1095
Age oldest trade	1	0.0434	0.00295	216.5251	<0.0001	0.0317
Age	1	0.0225	0.00222	103.2408	<0.0001	0.0257
Bankruptcies	1	−0.00342	0.000914	14.0060	0.0002	−0.00822

In the current example, I asked the students to provide *only* the core information about which variables were important to classifying good and bad credit risks—no betas, no *p*-values, no log of odds… just the variables. They were able to produce a graphic like that seen in Figure 3.8.

The graphic in Figure 3.8 is a depiction of the standardized coefficients. Because these values are standardized, the visual provides a simple comparison of relative importance of each piece of information to the model's ability to classify good and bad credit risks. While there is a great deal of detail that supports this graphic, the primary

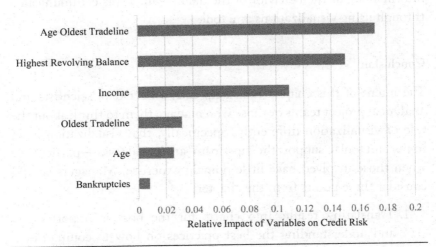

Figure 3.8 Graphic of relative impact on risk classification.

Table 3.4 Comparison of Risk-Modeling Results

CONTEXT	PROFIT PER 1,000 ACCOUNTS
Model results without visualizations informing the data preparation and cleaning process	$38,500
Model results with visualizations informing the data preparation and cleaning process	$92,250

message is that the *Age of the Oldest Tradeline* is about 50% more impactful on credit risk relative to the customer's income. In addition, *Income* is about four times more impactful than the *Age of the Oldest Tradeline* or the *Age of the Cardholder*. The details are important to keep as a backup to support the presentation if asked for by the client.

In the context of this project, given what they now know, I asked these students to go back and determine the profitability that their models would have generated per 1,000 accounts scored if they had not used the visualization tools to determine the behavior of the data. To calculate profitability, the students "won" $250 for every customer who they predicted as a good credit risk who were actually good credit risks, but "lost" half of the assigned credit line—average line was about $1,500—for every customer who they predicted as a good credit risk but actually defaulted.

The results in Table 3.4 are estimates and true results when implemented in the business may vary. However, there is no doubt that the results were improved by having the engaged analysts be more informed about the behavior of the data—which was communicated through using visualizations as a tool.

Conclusion

The intent of this chapter is to help early career data scientists and analytical project teams or those who manage them to think about the role of visualization differently. Specifically, that visualization is an important tool to support the "upstream" analytical tasks—particularly when those involved have little domain expertise. Managers should consider these points from the chapter:

1. Translating results into charts for the boss is important, and understanding the best practices on how to communicate results is foundational to the last phase of any analytical

project. However, this is merely one application of visualization, the simpler one.

2. Data scientists, computer scientists, mathematicians, and other computationally strong early career analysts frequently represent the paradox of visualization—as the individuals who can engage in the complexity of data analysis at a level that most cannot, they are more likely to discount the value that visualizations can add to their analysis, compared to analysts with less developed computational skills.

3. The ability to translate massive amounts of structured and unstructured data into information is in great demand across all sectors of the economy and across all application domains. The core skills needed for this translation are the same whether the area of application is health care, finance, retail, engineering, etc. As a result, when recruiting analytical talent, organizations often relax requirements for domain knowledge if they can obtain deep analytical skill. These analysts with deep skills, but limited domain knowledge, will be less likely to understand how data in a particular context are supposed to behave—particularly early career analysts. These individuals need to lean into the informational value that is inherent in the simplistic visualizations like histograms.

4. The creation of macros to simplify tedious and repetitive processes is a way that computationally strong individuals "show off"—it is often a game or a puzzle for them. The danger of this quick "macrotizing" of processes is that if they do not take the time to understand the underlying behavior of the data through visualizations, particularly if they have limited domain skills, their macros will be computationally impressive, but perhaps wrong.

4

MARKETING MODELS

Demonstrating Effectiveness to Clients

GREGG WELDON

Analytics IQ

Contents

The marketing industry is undergoing incredible changes, due to the continued increase in digital advertising. Traditionally, marketing was dominated by direct mail. Direct mail marketers purchase names (or lists of names) of consumers most likely to be interested in the product they're selling, craft a print advertisement for those consumers, and mail them out to households across America. It may take 4–6 months for this process to be completed. In addition, there is a delay of several weeks before the marketers learn the results of the mail campaign. Plus, the printing and postage costs of direct mail campaigns can be prohibitive. Hence, companies have a limited number of campaigns that can be done within a calendar year.

With the evolution of the Internet, targeted advertising has become an increasingly utilized method of getting a company's product out in front of the consumer. Online targeted advertising may take the form of e-mail marketing, banner ads, retargeting, search engine optimization, and addressable TV. Each method has its own strengths and weaknesses, but for this article, we'll focus primarily on display advertising on websites, the most common of which are banner ads. Those advertisements that pop up whenever you go to your favorite site? Those aren't random. Digital marketers can utilize cookies (small pieces of information in text format that are downloaded to your computer when you visit many websites) to "push" specific advertisements to specific people, much like direct mail marketers mail advertisements to specific homes.

With display advertising, the turnaround time is much faster and cheaper, as these ads can be sent to millions of people with a click of a button. Also, results from these ads can be compiled much faster than traditional direct mail campaigns. However, because of the fast and frequent nature of these digital campaigns, it can be difficult to get concrete results as to which ad campaigns were actually the ones that were the "deciding factor" in attracting new customers. If a person is hit with multiple advertisements in multiple formats from multiple Internet sites, which one(s) was (were) instrumental in capturing that consumer, and which ones were wasted effort? Another problem currently faced by display advertisers are the privacy concerns of matching cookies on a computer back to the personally identifiable information of an individual or household. Questions like these will take time to be ironed out.

As it exists now, many companies use both direct mail and display advertising to find new consumers of their products. The key takeaway for the data analyst is that, no matter what, the client needs good, solid information with which they can make their marketing decisions. How do we, as analysts, convey the value of our analyses to these clients?

Custom versus Generic Data Fields

Although the U.S. conducts a very accurate and complete census on every individual every 10 years, this information is compiled at the census block level, not the individual level. Breaking that down to a zip code or zip+4 level means making some very general assumptions.

Further breaking it down to the household level requires making even more assumptions. Basically, it's very difficult to "know" what is going on under every roof in America, despite some very good area-level data. Modeling companies attempt to take the area-level, or geographic, data that are available and find a way to model it down to the household or individual, without losing too much accuracy or predictability in the process.

In addition, many data companies specialize in compiling specific types of geographic, household, and/or individual data, such as ethnic information, gender information, auto ownership data, and retail transaction data. These sources can also be helpful in predicting the behaviors, attitudes, beliefs, and needs of individuals. Information compiled from nationwide surveys about every subject imaginable and reports on econometric changes across areas of the United States can also be used to give a clearer picture of people. There are literally hundreds of data sources available for the data analyst. Deciding which of these data sources are useful (many are garbage) and which ones are affordable (many are shockingly expensive!) can be a challenge in itself.

When building predictive statistical models, it's important to know whether the product you're working on is a custom model for a specific client or whether it's a generic product that your company hopes to package up and sell to a wide variety of clients after it's complete. Although the modeling techniques will be the same either way, how the model results are interpreted and conveyed will be vastly different. These differences will be detailed in the next sections.

Generic Model Visualization

A generic model is a project in which there's no specific client who has contracted for the work. Rather, the product is being built on the hope that many different clients will be interested enough in the final result to purchase the data field for use in their own business. For example, practically all modeling companies in the marketing industry have their own version of an income predictor, an age predictor, and a homeowner predictor. These fields have been developed using many of the types of data sources listed above and are designed by these data companies to give their clients the most accurate picture of consumers as possible.

Data companies that specialize in generic models like these are constantly seeking ways to make their predictive models more accurate. In order to achieve improved accuracy, there's sometimes a tendency to overfit models to the available data. If a model predicts that John Smith makes $50,124 a year in income, and he actually makes $50,124 a year, that's incredible (also, highly unlikely!). The downside to that level of amazing accuracy on John Smith, however, means that the same model may predict that Mary Jones makes $25,324 a year, when she actually makes $78,234. The cost of a high degree of accuracy for some individuals may be wild swings in predictions for others.

To avoid the overfitting trap, models must be validated with a holdout sample. Also, in practice, point estimates are less useful than a range (a confidence interval) within which some value is expected to fall. For example, we may predict that John Smith makes around $45,000–$55,000 a year, and Mary Jones makes around $75,000–$85,000 a year. Here, we lose some pinpoint accuracy, but we have a solid range of values for the majority of our observations. The goal of any data analyst is to improve both accuracy and reliability of the model, without sacrificing too much of either.

To put it in poker terminology, you might correctly predict that an opponent is holding *exactly* an A–J suited in the current hand, but you get bluffed completely on the next five hands. It may be more useful if you know the general range of hands your opponent is holding ("two high cards or a pocket pair") this time, as well as on an ongoing basis.

GamerIQ—A Generic Model

We have a new product at AnalyticsIQ called GamerIQ, which predicts the likelihood that a consumer is a video game player. This model was created as a direct result of seeing a need for it in the digital marketing industry. Several gaming companies were seeking information on gamers, and our company decided to attempt to fill that need. We conducted a national survey, using a professional survey firm that specializes in these types of analytics and identified people who considered themselves gamers. Our model was built using regular logistic regression, and we had a model with a score range of 0–999.

Surveys can be a challenge to model or, rather, to validate. We build the model on a survey dataset. Then, we ask the exact same questions to a different random population one to two quarters later and validate the original results.

Asking very specific questions like "Do you own a dog?" or "Do you own a smart phone?" generally validates well. However, asking a question such as "Do you *like* dogs?" or "Are you a sports fanatic?" is nearly impossible to validate. Asking questions about peoples' feelings can be hard to validate, because the answers we get can vary widely, based on a variety of factors. "Are you a sports fanatic?" Well, that would depend on each person's image of what a sports fanatic looks like. What if a person loves baseball more than life itself, but we ask the question during the football season? Will his answer be different? What if the subject answers the question on a cold, rainy day, versus answering the same exact survey question on a beautiful Saturday afternoon? All of these factors (and many, many more) can affect survey results on "feeling" questions like these. For accuracy and predictability on these subjects, we rely on a series of techniques that are beyond the scope of this article. Building stable models on feelings is like trying to catch a greased pig—very tough and messy!

GamerIQ was built and validated well. However, unlike data analysts, most clients in the marketing industry don't want to try to interpret a score ranging from 0 to 999, which is typical for many predictive models. In recent years, we've moved to using a 7-point scale, where "7" represents "Definitely Yes" and "1" represents "Definitely No" in most cases. This type of scale gives the client the information they need without a lot of clutter.

For this model, however, we decided to become even more "client friendly" by researching the video game industry further. The industry breaks up gamers into four categories: Hardcore, Core, Mid-Core, and Casual players. We elected to follow suit. Since the industry estimates that 2% of their players are considered Hardcore, we took the top 2% of scores on our model and made them a "7" (Hardcore). We did the same with Core ("6"), which is roughly 4% of players, Mid-Core (10% of players), and Casual (32% of players). We also know that around 50% of Americans don't play video games at all, so the bottom scoring 50% of our scores were given a value of "1."

What about scores of "2" through "5"? We wanted to try to further segment the Mid-Core and Casual gamers; there were enough people in these two groups for statistically significant segments, and it could serve as a differentiator between our gamer model and any gamer models that our competitors may eventually build. We tried several types of splits, such as families versus individual players, but nothing really stood out until we tried gender. Male versus female gave us the unexpected bonus of identifying two underserved groups in the video game industry! Table 4.1 shows the means analysis that we created while analyzing our data.

These types of tables are invaluable for identifying groups that are unique and different. Along the top of the page, you can see that we have our total sample (this is a random 1% sample of the U.S. adult population, on which we applied our GamerIQ model), followed by our seven groups, ranging from Hardcore to No Interest. We include our sample size and percentage, our average GamerIQ scores, as well as a series of means for various data fields that can be used to summarize each group.

For example, Hardcore gamers are 2.01% of our sample and had an average GamerIQ score of 458. Their average age is 32.41, with 12.64% of them in the 18–22 age range (highest among the groups). Highlights include their annual income on average $51,870 (IncomeIQ_Plus_v3), and their home value is $208,260 (HomeValueIQ). They use Facebook at a higher rate than any other group (5.87—SocialIQ_Facebook_v2), and they're least likely to be married (AIQ_Marital—Married) and have the lowest educations (AIQ_Education—highest "Less Than High School" and lowest "Bachelor Degree" and "Graduate Degree"). Take a look at EthnicIQ_v2 and AIQ_Gender; the Hardcore group is almost exclusively male, but is by far the most ethnically diverse.

Selling the Model

The two surprises from this chart are Group 4 (Mid-Core Female) and Group 3 (Casual Male). Mid-Core Females make up 2.43% of the total population. They're younger, have lower educations, have more children in the home, and have less education than average. Casual Males on average 65 years old (!) make up 11% of the population,

Table 4.1 GamerIQ Means Analysis

		7.00	6.00	5.00	4.00	3.00	2.00	1.00
	TOTAL	HARDCORE	CORE	MID-CORE MALE	MID-CORE FEMALE	CASUAL MALE	CASUAL FEMALE	NO INTEREST
Sample size	N	2,142,613	42,968	84,073	132,462	52,145	240,077	456,628	1,120,748
	%	100.00	2.01	3.92	6.18	2.43	11.20	21.31	52.31
SCR_Gamer	M	116.59	458.28	378.80	297.32	251.51	128.21	110.85	54.87
	min	22	400	342	209	209	100	50	22
	max	673	673	399	341	341	208	208	99
AIQ_Age (aged 18–22)	M	53.07	32.41	32.48	33.47	33.55	65.21	48.27	59.83
	%	2.88	12.64	11.35	9.31	8.06	0.00	4.75	0.00
OS_Innovator	M	4.04	6.27	6.25	5.78	4.75	4.35	3.51	3.71
IncomeIQ_Plus_v3	M	82.60	51.87	73.70	105.42	51.25	44.62	71.70	95.86
WealthIQ_Plus_v4	M	–	140.19	242.27	585.42	143.11	187.19	338.70	574.84
Spendex_Plus	M	10,384.74	7,766.43	10,183.28	14,816.35	7,507.16	4,866.37	9,276.09	11,753.87
HomevalueIQ	M	295.44	208.26	256.45	323.53	200.51	199.33	253.91	338.50
SocialIQ_Facebook_v2	M	4.14	5.87	5.37	3.89	5.77	4.84	4.41	3.67
SocialIQ_Twitter	M	3.63	4.73	4.91	5.09	4.13	2.53	3.33	3.65
SocialIQ_Instagram_v2	M	4.05	5.38	5.21	4.63	5.55	3.31	4.36	3.80
AIQ_Marital									
Married	%	45.14	23.02	31.43	42.58	29.32	43.00	45.25	48.28
Single	%	29	64.52	54.53	42.77	58.79	23.08	37.17	20.60
Unknown	%	22.95	12.45	14.05	14.42	11.89	28.06	14.14	28.25

(*Continued*)

Table 4.1 (*Continued*) GamerIQ Means Analysis

		TOTAL	7.00 HARDCORE	6.00 CORE	5.00 MID-CORE MALE	4.00 MID-CORE FEMALE	3.00 CASUAL MALE	2.00 CASUAL FEMALE	1.00 NO INTEREST
EthnicIQ_v2									
AA	%	10.47	34.25	14.33	3.64	31.20	17.76	12.95	6.56
Asian	%	3.37	3.82	4.22	2.89	2.52	3.12	2.30	3.84
Caucasian	%	67.75	39.76	58.40	77.31	45.34	59.92	67.57	71.17
Hispanic	%	13.35	19.53	18.99	10.12	18.47	15.57	13.13	12.47
Other	%	5.06	2.63	4.06	6.04	2.45	3.59	3.64	5.96
Presence_of_Children									
No	%	19.68	8.13	8.58	11.16	6.97	26.53	19.36	21.15
Yes	%	35.54	36.33	40.09	45.77	51.75	26.49	42.41	32.21
Unknown	%	44.78	55.54	51.33	43.07	41.28	46.98	38.23	46.64
AIQ_Gender									
Female	%	49.54	0.64	2.03	–	100.00	–	100.00	49.14
Male	%	48.74	93.58	93.37	100.00	–	100.00	–	49.35
Unknown	%	1.72	5.78	4.60	–	–	–	–	1.51
AIQ_Education									
Less than HS	%	5.64	14.89	9.09	2.61	12.45	9.58	5.83	4.15
HS degree	%	46.15	50.65	49.32	46.95	52.44	54.37	50.11	41.90
Bachelor degree	%	27.19	18.33	21.71	30.46	20.27	21.06	26.40	29.53
Graduate degree	%	8.8	3.12	4.05	7.02	3.31	4.74	5.53	12.08
Other	%	12.22	13.02	15.83	12.96	11.54	10.24	12.13	12.34

have low incomes (probably many retirees), have few children in the home, and are racially diverse. These two groups of people play video games, but are not groups that have traditionally been targeted for video game advertisements. Would a specialized series of advertisements, focused on these two groups, move the needle in video game sales in certain circumstances? That's the opportunity that our clients may want to explore.

Having said that there are a lot of numbers on Table 4.1, and, as stated before, many clients don't have the time or inclination to review pages and pages of Excel spreadsheets. Figure 4.1 shows a series of charts that were put together, based on the numbers from Table 4.1.

Here, you can clearly see the contrasts between the groups in terms of age, income, home value, social media interests, ethnic groups, education levels, and basic demographics. These are charts that are routinely placed into PowerPoint presentations for client calls. In addition, customized charts can be created on the fly for any specific requests that a potential client may have.

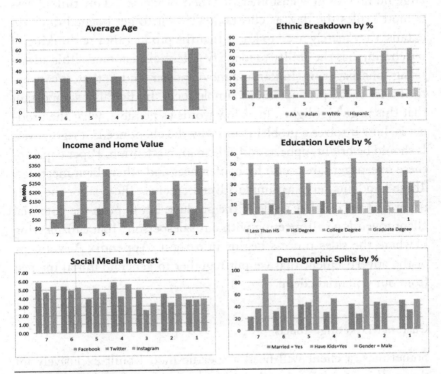

Figure 4.1 GamerIQ charts.

For example, Table 4.1 clearly shows the differences in ages by groups, showing that Hardcore, Core, and Mid-Core players are much younger (on average) than the Casual and No Interest groups. Likewise, differences in ethnic groups, income, social media interests, and education levels by group can be more readily identified by the client as potential targets for marketing segments and sales efforts. A cursory look at the charts tells us that Group 3 (Casual Males) are older, have lower incomes (many are retired), have fewer children in the home, and are the least likely to use Twitter and Instagram. These players don't respond to the same marketing efforts that work with more traditional players. A client interested in reaching out to these potential customers may elect to contact them through direct mail, television advertising, and even magazine ads.

For generic products, our sales people and consultants can use the information we prepare to get clients interested in a certain product. We can then create any charts and graphs that we need to help demonstrate that our product can solve their problems and identify a large number of new customers for their products. Many times, "less is more" rules the day, as clients don't want to get bogged down in details; they merely need to know that the product they're thinking of purchasing has the ability to identify potential customers. This advertising "one-pager" is similar to a movie poster, and in that, it conveys the basics of the product in one simple sheet. In digital, it's important that this page highlights, in as few words as possible, what the product is, how it solves a problem, and why the client needs it right away. It should be bright, upbeat, and able to grab attention. Paper copies of these one-pagers are useful for conferences and other client meetings. These handouts serve as simple advertisements. The goal of the one-pager isn't to complete a sale of the product, but to push the potential client towards a call/meeting with our salesperson. That call/meeting is where the sales process begins in earnest. **Appendix A** is a copy of the advertisement that we put together to generate initial interest from our potential clients.

Custom Marketing Models

Sometimes, a client will need a specific model built exclusively for them. This may be a situation where they need the results of some previous marketing campaign modeled in order to improve result for the next campaign; they may have a new product they want to introduce into the marketplace, and need some direction as to how it may perform. In some cases, the client may provide their own, internal data for use in the creation of the custom model.

In previous years, custom models were the bread-and-butter for modeling companies. However, as statistical modeling has increased in demand, more and more people are learning the business, and companies are able to staff their own internal modeling units. These modeling units need data, which in turn has given rise to the increase in generic models as we discussed in the last section.

Custom model clients require a different level of detailed information than we typically give generic model clients. In a custom model, we're giving our clients strategic instructions on how to maximize their results. We're using the model to forecast the actual rate of "success"; however, the project in question defines success (i.e., response, conversion, and attrition).

Table 4.2 is an example of a simple response model that was built for a direct mail client. We had a sample of 20,486 observations, and of which 16,122 did not respond and 4,364 did respond. As shown in the top section of the report, we have a forecast for how the resulting model would have performed (and, presumably, *will* perform when used in the next marketing campaign). The data have been broken into ten deciles and, reading across, show the total population, the nonresponders, the responders, the response rate, the response odds, and the Kolmogorov–Smirnov (KS) values.

Let's assume that we recommend to the client that they mail the top three deciles. This means that everyone that scores 250 or higher on the model will get mailed. The total mailed group will be 2056 + 2057 + 2275 = 6388 (see the column titled "Total Population—Records"). This represents 31.2% of the total population. Reading across the line "Decile 3" shows that we would be mailing to 28.0% of the nonresponders but 43.0% of the responders. Under "Response Rate-Cum. %", our response rate will be 29.37%, giving us a "Response Odds Cum %" of 1.38. This translates into a roughly 38% increase in response rates from the client's current results (current response rate = 21.30%; new response rate will be 29.37%).

Table 4.2 Client Response Model

DECILE	SCORE	TOTAL POPULATION RECORDS	TOTAL POPULATION INT. %	TOTAL POPULATION CUM. %	NONRESPONDERS—BAD RECORDS	NONRESPONDERS—BAD INT. %	NONRESPONDERS—BAD CUM. %	RESPONDERS—GOOD RECORDS	RESPONDERS—GOOD INT. %	RESPONDERS—GOOD CUM. %	RESPONSE RATE INT. %	RESPONSE RATE CUM. %	RESPONSE RATE DECUM. %	RESPONSE ODDS INT. %	RESPONSE ODDS CUM. %	RESPONSE ODDS DECUM. %	KS (%)
1	308–High	2056	10.0	10.0	1362	8.4	8.4	694	15.9	15.9	33.75	33.75	21.30	1.58	1.58	1.00	7.5
2	273–307	2057	10.0	20.1	1498	9.3	17.7	559	12.8	28.7	27.18	30.46	19.91	1.28	1.43	0.93	11.0
3	250–272	2275	11.1	31.2	1652	10.2	28.0	623	14.3	43.0	27.38	29.37	19.00	1.29	1.38	0.89	15.0
4	226–249	2242	10.9	42.1	1703	10.6	38.5	539	12.4	55.3	24.04	27.98	17.65	1.13	1.31	0.83	16.8
5	203–225	1879	9.2	51.3	1503	9.3	47.9	376	8.6	64.0	20.01	26.56	16.44	0.94	1.25	0.77	16.1
6	186–202	1835	9.0	60.3	1479	9.2	57.0	356	8.2	72.1	19.40	25.49	15.77	0.91	1.20	0.74	15.1
7	172–185	2004	9.8	70.0	1651	10.2	67.3	353	8.1	80.2	17.61	24.39	14.95	0.83	1.15	0.70	12.9
8	149–171	2051	10.0	80.0	1745	10.8	78.1	306	7.0	87.2	14.92	23.21	14.08	0.70	1.09	0.66	9.1
9	125–148	2048	10.0	90.0	1744	10.8	88.9	304	7.0	94.2	14.84	22.28	13.65	0.70	1.05	0.64	5.3
10	Low–124	2039	10.0	100.0	1785	11.1	100.0	254	5.8	100.0	12.46	21.30	12.46	0.58	1.00	0.58	0.0
	Total	20486			16122			4364			21.30			1.00			

(*Continued*)

Table 4.2 (Continued) Client Response Model

VARIABLE	POINT VALUE	% OF POP	RESPONSE RATIO	POWER RANKING	VARIABLE RANGE
Intercept	####	100.0	1.00		
AIQ_ADDRESS_INDICATOR	0.3856	10.1	1.33	1	= 0–180
		29.9	1.18		= 181–348
GC_MTTRDS	0.1632	30.8	1.23	8	= 0–0.64
AUTOPROP_GMC	0.1360	20.3	0.68	4	= 5–7
	####	14.7	1.40	3	= 0–1.4
E_MED_NUM_VEHICLES	0.3107	29.0	1.25	6	= 43.72 or greater
MARITAL_NEVER_PCT_V2	0.1884	9.7	1.34	5	= 0–40.38
RELPCT_NOUNK_V2	0.2761	10.0	0.74	2	= 89 or greater
Focus_PLUS	####	30.4	0.82	9	= 0–152
SCR_DYN_BOUNCE	####	10.0	0.72	7	= 4736 or greater
WEALTHIQ_MAX_PLUS	####				

AIQ_ADDRESS_INDICATOR	Likelihood that an address is real
AUTOPROP_GMC	Likelihood of purchasing a GMC vehicle (7 = most likely, 1 = least likely)
E_MED_NUM_VEHICLES	Median number of vehicles in the zip + 4
GC_MTTRDS	Average # of mortgage trades
MARITAL_NEVER_PCT_V2	% Never married
RELPCT_NOUNK_V2	% of population who are not members of any religious organization
FOCUS_PLUS	Payment propensity score
SCR_DYN_BOUNCE	AIQ Email deliverability model
WEALTHIQ_MAX_PLUS	High-value net worth predictor ($6MM+) (in $'000)

Proprietary & Confidential, AnalyticsIQ (2016).

Why choose to go down three deciles in this example? It's a combination of several factors. First, we want to mail enough offers to get a significant number of new customers. We could mail just the top 5% of scores, but the total number of new customers wouldn't be enough to justify paying to have the model done in the first place. Mailing all the way down to the seventh or eighth decile would bring in many new customers, but the client would be upset with the number of nonresponders and, by definition, number of wasted dollars in printing and postage costs. Each marketing campaign has a very specific budget, and every dollar spent on nonresponders on this campaign is a dollar less to spend on the next campaign.

Ideally, we can find the sweet spot where the client is bringing in enough new customers to make the campaign worthwhile, while also limiting the number of mailings that go directly into someone's trashcan.

This top section of Table 4.2 should be studied in great detail, as the success of the project will be judged on how well actual results line up with this forecast. Marketing clients are mainly concerned with how much of an improvement in response rates they'll see with the new model. A cumulative response rate of 29.37% that we've forecasted above is what they're going to use to project future revenues and budgets over the next fiscal period. Some deviations from this forecast are to be expected, of course, but major differences can scuttle the project (as well as the client relationship). Thus, the model's ability to improve over current results is the key factor in every marketing project.

The second section of Table 4.2 is the model itself. Since this is a custom model, the client has a right to see what variables are being used in the model. In some cases, clients may not want certain variables used in their model. These may be variables that could be problematic from a regulatory perspective, variables that have added costs associated with them, variables that seem counter-intuitive to the client, or, sometimes, variables the client just doesn't like. Here, we're showing the name of the variable, the Parameter Estimate from the regression itself, the percentage of the population that is affected by that variable, the Response Ratio (positive or negative), a Power Ranking, and the Variable Range (when using dummy variables).

The Response Ratio is a nice way of showing the client, in a very straightforward way, how each variable affects the score, positive or negative. Many times, clients will be surprised when a variable skews one way, when they were convinced it "should" be the opposite. This is a great time to uncover these kinds of inconsistencies and discuss them. Sometimes, the client will yield to the data; sometimes, the variable is jettisoned. The Power Ranking is an informational field that shows the order in which each variable varies from the neutral or zero. Variables with the potential to move a person's score the most, up or down, have a higher Power Ranking than a variable whose Parameter Estimate is approaching zero.

The bottom of Table 4.2 shows a definition of each variable in the model. Further, more detailed information for each variable is available if requested.

Obviously, the information provided for custom models is much more detailed than for generic data fields. The main reason is that custom models belong to the client and are being built at their express direction in order to solve a specific problem. This information is required for setting strategies and, after the campaign is completed, for validation and follow-up. Meetings with clients to go over these reports are generally long and cover many topics in great detail, so that, in the end, the client has a comfort factor with the model that they will be using to command their next marketing campaign.

Do's and Don'ts in Client Meetings

Many times, a meeting with a client will entail either a face-to-face discussion or simply a conference call to review the results. These interactions will be different, based on if the product being discussed is a generic product or a custom product.

For generic products, the meeting/call will primarily be of a sales nature. You have a product that has presumably been reviewed by your company's sales staff, and an "elevator pitch" has been created for it. This elevator pitch is simply a short, concise explanation of what the product is and how it can solve the potential client's problem. It should be clear, simple, and able to convey all the basic information in the time it would take to ride an elevator with that client from one floor to the next. You could also view those short, 15-second TV commercials as elevator pitches.

As a data analyst, your work has already been completed. You've presumably built an accurate and/or predictive product that validates well and is in demand by a fair number of potential users. You've completed statistical analysis that shows how well the model works on the intended consumer base, and created supporting documentation that the sales and marketing staff of your company have used to create their sales presentation.

If you're included in the client meeting, you are generally there as backup in case the potential client asks questions that no one has anticipated (this usually happens when the product being discussed is still new and untested). As the resident expert, your job is to speak intelligently about how the product was built without going into too much detail about the actual variables used or techniques employed. Remember, this is a generic product, and clients are more interested in how the product will make them money than all the fascinating details about your first regression correlation matrix! You'll want to follow the sales person's lead in these meetings. Let them guide the client through the sales process and only add in specific modeling details when appropriate.

Many times, clients may be unsure of a new product, especially if it involves something they haven't purchased before. Clients may represent large corporations, but the people you're talking to are just individuals. Your client contact won't get fired if he/she purchases the same old products that they always have in the past. However, they *could* be fired if they take a chance and purchase your brand new product and it doesn't perform up to expectations. Always keep that in mind in these meetings.

Many clients will be less interested in the details of the regression than in getting a "positive feeling" about the people (i.e., you) who built the product. If you can convey intelligence, confidence, and enthusiasm in the meeting/call, many clients will feel more comfortable in the product, in your company, and in you, and will be more likely to purchase the product.

Custom model meetings, on the other hand, will be much more detail oriented. These clients have contracted with your company to build a specific model for their specific needs, and they'll be highly interested in the modeling techniques used, any challenges or data problems that had to be overcome, and how the results of the model will specifically be used to make them money. Be prepared for these meetings by reviewing all of your notes and programs from model development in advance. If needed, prepare accompanying documents and charts, and send them to the client a day or two in advance, so that everyone in the meeting will have the information at their fingertips.

You may be called upon to give a summary of each step of the process. Have this summary prepared in advance, either on notecards or memorized, so you can speak concisely and intelligently. The ability of the analyst to speak with confidence in the meeting and answer any questions that may arise goes a long way in getting a sale closed.

Enthusiasm can cover a lot of flaws when presenting a subject. If a person who is presenting a subject is upbeat, positive, and enthusiastic about the work, the product, and, to a certain extent, just being in the room talking to the client, it makes those clients much more at ease. Many deals have been consummated with a client saying, "Well, I don't know what that guy was talking about, but *he* certainly did!" Believe it or not, that can be the tiebreaker in many situations.

Be prepared for as many eventualities as possible before going in to the meeting by reviewing the project and practicing your presentation prior to the meeting. Follow the salesperson's lead. Discuss your portion of the project with knowledge and enthusiasm.

Be ready to alter your presentation on the fly if you "read the room" and realize that the client isn't really interested in what you were about to discuss. If the client seems bored or distracted during your presentation, you may be getting too detailed for them; switch to a more general approach to get them reengaged. If the client is ready to buy, stop "selling" and take their money!

Conclusion

Data visualization can be almost as important as the development of the product itself. Many times, the way the results are presented to the potential client is critical in whether the product is ever sold. The product may be the best in its class, but if that information isn't conveyed in a clear, concise, and pleasing manner, it may be doomed to failure. Many products that are completely terrible make a lot of money each year, thanks to the expert marketing efforts of a dedicated group of people intent on selling to as many clients as possible. It can be a tough lesson to learn when your highly predictive, expertly crafted product is continually beaten in the market by a slipshod mess that had a great marketing campaign behind it.

Once your product is complete, try a lot of different techniques to show off what you've created. Ask questions. Find out from clients what kinds of information they're looking for in order to make a purchasing decision. Find out from friends or family members, people not familiar with your industry, which visualization techniques are most understandable to them. As data analysts, we may feel right at home with stacks of statistical analysis software (SAS) or R program output and simple KS charts. Remember, most people don't feel the same way. In order for our product to succeed in the marketplace, we need to rely on people who don't have our skillsets, but have other, just as important, skills. It's a team effort.

Appendix A: Game Over—AnalyticsIQ Is Proud To Release GamerIQ

AnalyticsIQ is a marketing analytics firm that builds powerful and predictive consumer data products that fuel analytical performance for data-driven marketers.

Let's Play

According to the Entertainment Software Association, a trade group for the video game industry, 59% of all Americans play video games, nearly half of those who play are women, and the average age of a gamer is 31. While these facts may be surprising, the video game industry has quietly attracted many newcomers who have not traditionally been game players thanks to new technologies and diverse offerings.

With record-breaking revenues realized nearly annually, the expanding market for all things video games shows no signs of slowing down. But the wide range of game genres, the varying degrees of complexity and commitment, and distinct consumer tastes make targeted advertising particularly challenging for video game marketers.

How then can they get "one-up" on the competition to profit from unprecedented demand?

GamerIQ

Meet GamerIQ, AnalyticsIQ's latest release of "gamer" audience segments. GamerIQ accurately identifies not only individuals who have an affinity for gaming but also the level of interest and enthusiasm the consumer has for video games. In fact, AnalyticsIQ (AIQ) has created six audience groups based on the industry-standard gamer dedication spectrum. Categories include the following:

- Hardcore gamers
- Core gamers
- Mid-core male gamers
- Mid-core female gamers
- Casual male gamers
- Casual female gamers

GamerIQ was developed based on proprietary survey responses to questions crafted by an on-staff cognitive psychologists and leverages consumer motivations as well as predictive lifestyle, demographic, and finance data to give video game marketers unparalleled insight into their target market.

Level Up

As competition surges, video game marketers need every available advantage to attract and acquire new customers. GamerIQ can take video game marketing to the next level by giving marketers the ability to

- Lower marketing expenses by effectively targeting likely gamers.
- Match prospects with the best offers based on interest and dedication level.
- Increase individual consumer wallet share.

How AIQ Data Compares to Other Providers

- 15%–25% more predictive
- 11% higher coverage of U.S. households
- 12%–22% more accurate

5

RESTAURANT MANAGEMENT
Convincing Management to Change

WILLIAM SWART
East Carolina University

Contents

Introduction

The Pillsbury Restaurant Group included Burger King, Steak & Ale, Bennigan's, Godfather's Pizza, and Haagen-Dazs. The focus of this chapter is on Burger King, the first restaurant company to adopt analytics in the form of Operations Research and Industrial Engineering. The setting is the heights of the hamburger wars during the mid-1970s and mid-1980s when Burger King was aggressively attempting to take market share from the market leader McDonald's. Although Burger King was founded in 1953, two years before McDonald's, it found itself lagging as consumer preferences for fast food changed in increasingly health conscious societies in which more families opted to eat out than ever before.

Although Burger King and McDonald's appeared as "burger joints" to the casual observer, they were very different businesses. McDonald's was a hamburger business in the true sense of the word. They owned roughly 80% of their restaurants and maintained tight QSCV (Quality, Service, Cleanliness, and Value) standards through a management hierarchy that stretched uninterruptedly from store to corporate headquarters. Burger King, which had been acquired by Pillsbury Corporation in 1967, was a financial company with 80% of their stores being franchised. Thus, while McDonald's sold hamburgers to make a profit, Burger King sold franchises.

In addition to being subservient to a parent company whose main line of business was not restaurants, Burger King management had to maintain a productive relationship with their franchisees. The guiding principle of franchising is that what is good for the franchisor must also be good for the franchisees, who pay a certain percentage of their revenues to the franchisor. Franchises are independent businesses within the Burger King system, and the most significant power the franchisor has is its ability to not renew the franchise agreement when it expires. As a result, the franchisor and franchisees are constantly seeking an amicable balance between what is good for one and the other.

This amicable balance is maintained through Burger King being able to demonstrate that their business model (products, operations, marketing) is best. The principal strategy for doing this is to demonstrate that the business model is the best for the substantial cadre (20%) of company-owned stores. But, there are always franchisees who have a "better" idea, and their view must be given due consideration if a harmonious relationship is to be maintained. Being able to do so requires the ability to objectively evaluate ideas and an effective way to communicate the results of the evaluation.

Strategy and Operations

The single determinant of restaurant sales capacity is speed of service—the time it takes between when the customer places and receives the completed order. The faster the speed of service, the more people can be served; therefore, the higher the restaurant's potential sales volume. Typically, for every dollar a Burger King restaurant increases its lunch hour sales, it will gain four times as much during the day.

Burger King's strategies to increase its market share could only be implemented if the restaurants could service the projected increases in customer counts while maintaining a "sales building posture." This meant that even at the busiest time, customers could be served within the QSCV standards set by the corporation. In the face of increasing customer counts, a sales building posture could only be achieved by adding labor hours to the restaurant, developing more efficient manufacturing systems to handle product customization and diversification, or both. Since adding labor to an inefficient manufacturing system creates a permanent stream of unnecessary costs, a more efficient restaurant had to be designed capable of accommodating new products as well as the "Have it Your Way" promise made to customers. Simultaneously, a productivity improvement program had to be developed to upgrade existing restaurants.

Finding and Assessing Performance Improvement Projects

In a largely franchised system like Burger King, there is never a paucity of improvement ideas put forward by the franchise community as to how things should be done differently than what the corporate standards specify. Burger King Corporate responsibility was to vet ideas put forth, as well as add their own ideas, and then seek to build a consensus among franchisees as to what should and should not become part of the Burger King Operations Manual. The improvement ideas could include everything from menu items to dining room décor standards, equipment standards, and layouts for the kitchens, dining rooms, and drive-thru lanes.

These combined ideas of franchisees and corporate personnel provided a continuous cycle of potential projects for performance improvement. A systematic and controlled process to manage relevant and appropriate projects for sustainable implementation was required. A coauthor and I proposed a four-phase model, similar to the process used at Burger King, for managing the introduction of performance improvement projects to insure the best and brightest ideas are implemented and those that are discarded along the analysis path are rejected for good and coherent reasons. The key step in the first phase is to analyze the impact of the project to determine it has any impact on operations, on speed of service, or on restaurant profitability.

Before and early into the hamburger wars, the Marketing Research Department performed the analyses of the impacts of operational changes on speed of service. At the time, the Marketing Research Department had three groups: Consumer Research, Sensory Research, and Operations Research. The Operations Research Group was responsible for determining the impact of any operational change on speed of service. Note that at the time, Operations Research at Burger King meant conducting speed of service studies at the store level—not Operations Research as a separate discipline.

The process followed in a speed of service study was time consuming, laborious, and costly. Burger King would send analysts equipped with forms and stopwatches to record the actual speed of service at their six R&D restaurants located near corporate headquarters. After an adequate sample had been obtained, the proposed operational change would be physically built into the R&D restaurants, and the crews would be trained so that the modified R&D restaurants could be put into operation. After a suitable period, the analyst would again record actual speed of services. The difference in pre- and postchange in speed of service would then be statistically analyzed to determine if the difference, if any, was statistically significant. Simultaneously, if appropriate and warranted, members from the Consumer Research Group would assess the impact of the change on customer perceptions and, if the change involved a new product, members from the Sensory Research Group would assess customer taste.

Communicating Results

Top management at Burger King consisted mostly of individuals in their mid-30s. Even Burger King's current CEO, Daniel Schwartz, is 36 years old. Their age belied their experience. Most had started with the company as teenagers flipping burgers and the like. Burger King was expanding at the rate of one restaurant per day, and every six restaurants required a company District Manager. Successful Company District Managers were rapidly promoted to Franchise District Managers who, in turn, were promoted to Area Managers and then to Regional Vice Presidents. While the wages of a restaurant crew member were low, management jobs paid very well considering the young age and lack of formal education of most incumbents.

Candidates selected for promotion were those that could size up a situation, make a quick decision, and then implement it successfully. There was no time or capability for collecting and formally analyzing data. Those that rose to the top constituted a bureaucracy of bright, young, and aggressive entrepreneurs who neither asked nor gave quarter. Success was quickly rewarded, and failure was fatal as demonstrated by the fact that during the heights of the hamburger wars (1978–1986), Burger King had a new CEO every 2 years.

The counterbalance that tempered the top-down management style of the corporation were the franchisees. While not directly in the chain of command, they wielded substantial influence. For example, Chart House was Burger King's largest franchisee at the time with over 400 restaurants. Other franchisees were or had strong relationships with members of the board of directors of Pillsbury.

In order to seek support from the franchisee community, Burger King Corporate had to be able to show objectively that the performance improvements that they recommended were indeed an improvement and, above all, that they did not negatively impact speed of service. This, to a large degree, justified the time-consuming and expensive speed of service study methodology described in the previous section.

Top management at Burger King had no interest in reading long reports or being presented with methodologies that were used. Their basic attitude was that you were paid to obtain results. How you obtained them was not their concern. If the presentation was too long, they simply walked out (not a good omen for the presenter). Their total focus was on achieving positive results. They wanted to see the numbers and have their questions answered. Transparencies or slides were the media for communication.

In order to lay the groundwork for a successful presentation, the project team must have a champion within the management structure who was willing to endorse the project to the corporate structure. The project should also be advanced to subordinate leadership levels for consensus building. Consensus building allowed managers and the people ultimately responsible for implementing the project to express support or, if appropriate, to withhold their support. Even when a project had received endorsement and consensus had been built for its implementation, there may still be concerns within the organization regarding its advancement. It was a good idea to make every effort to bring these concerns to light and discuss these openly before the presentation. Having a presentation lapse into a discussion of concerns would not lead to the desired action.

Once the groundwork described above had been laid, the project team was ready to focus on the presentation itself. Each presentation had a different audience and each member of the audience had

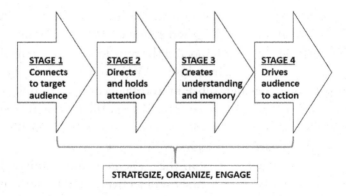

Figure 5.1   Anatomy of a presentation to burger king corporate management.

different responsibilities and/or priorities. As shown in Figure 5.1, presentations were carefully strategized and organized to engage management during each of the four conceptual stages of the presentation. Stage 1 starts the presentation by clearly stating the business issue that the presentation will address. Stage 2 directs the presentation to identifying what is in it for each stakeholder in the room. Stage 3 emphasizes what the benefits of implementation of the project will be, and Stage 4 will specifically state the actions that each stakeholder in the room must commit to in order to achieve the projected benefits. The target time for the entire presentation was 30 minutes. All presentations were straight and to the point, and drama and acting had no place in the presentations.

From Marketing Research to Operations Research

Early during the burger wars, a position of Operations Research Analyst in the Marketing Research Department became vacant. The vacant position was advertised. One of the applicants for the position was a candidate that had obtained a MS degree in Operations Research. During this person's interview, Burger King first became aware of the existence of an academic degree in Operations Research. At the same time, the candidate became aware that what Burger King was asking for was not exactly what he expected. During the interview, the Burger King interviewer asked the candidate whether he knew statistics. The response was an emphatic affirmative, and the candidate was hired.

Not long thereafter, the manager of Burger King's Operations Research Department was promoted to Director of Marketing Research. The recently hired analyst had impressed his superiors to the extent that they wanted to expand the scope of the department from just statistical analyses to include simulation and optimization. The new analyst was considered too inexperienced to be promoted to manager. However, he was asked to help identify a suitable manager.

The new analyst contacted one of his former professors (the author) to inquire about his interest in the position. Initially, the professor expressed dismay that student had accepted employment at a burger joint. However, after conversation with his former student as well as with Burger King Management, the professor realized that Burger King was virgin territory to Operations Research and that the position represented an enormous opportunity to introduce Operations Research not only to Burger King but also to the restaurant industry as a whole.

The newly refocused Operations Research Department was repositioned as an internal consulting function to the corporation. Consisting only of the manager and one analyst, it was charged with "selling" its services to whomever would "buy" them. This meant that any organization using the new department would have to transfer a portion of its budget to that department. In turn, the new department was required to produce enough budget transfer to "pay" for the analysts. The manager's salary was considered as corporate overhead.

This new department, the first of its kind in the industry, made immediate impacts on Burger King. It developed a hamburger formulation model that is credited to have saved Burger King $1 million/year in meat costs. It also used simulation as a tool to develop the double window drive-thru that is a common sight at many fast-food restaurants today. The double window drive-thru was one of the earlier undertakings of the Operations Research Department and provided valuable lessons on how presentations impact different members of the audience in different and sometimes unexpected ways.

Case Study 1—The Double Window Drive-Thru

1. What Was the Problem?

Up to that time, drive-thru customers would place an order at the order station and then join a single line, single server system to receive their order. The server would accept payment, make appropriate change, and then deliver the completed order to the customer, who would then drive away. As society became increasingly mobile, the popularity of drive-thru's grew to the point that almost 50% of a restaurant's daily sales were made at the drive-thru. It was not uncommon to see cars lined up to the street waiting to place their orders. It also was not uncommon to see cars that intended to use the drive-thru not join the line, or balk at the waiting times and leave the line. Regardless, a lost customer represents lost revenue and a customer not satisfied with long waiting times is less likely to return to the establishment in the future.

Many ideas were explored to increase the capacity of the drive-thru. Perhaps most noteworthy was an attempt championed by a very senior member of management to create parallel servers as used by many banks. Instead of vacuum tubes, conveyors were constructed to transport orders to the outside lanes. Unfortunately, the transport time to the outer lanes increased the customer waiting time and negatively impacted food quality. The cost of this failed experiment was in excess of $100,000.

The Operations Research Department took on the challenge of developing an approach to increase the drive-thru capacity. They knew that after the parallel server fiasco, Burger King would be very critical of any new ideas and reluctant to spend money on testing new ideas unless convincing evidence was presented that they would work.

2. What Analyses Were Performed?

An analysis of the current drive-thru system revealed that after customers received their change at the drive-thru window, they had to wait on their order to be "manufactured and assembled." The waiting time was attributed to the manufacturing system not having enough lead time between the time the order was placed and the customer was ready to receive it.

The Operations Research Department hypothesized that if the two activities of making change and waiting on the order could be disaggregated so that making change occurred before customers proceeded to pick up their order, then customers could be discharged from the system at a faster rate. For example, if making change took 9 seconds and waiting on the order took another 51 seconds, then customers would wait at the drive-thru window for 60 seconds. However, if customers could pay and receive change at an alternate location (9 seconds) before reaching the drive-thru window, then they would only have to wait 51 minutes at the window. This change from 60 to 51 seconds at the drive-thru window represented an increase of 15% in drive-thru sales capacity.

The problem that remained to be solved was where/how to make change before the drive-thru window. Working with Burger King's construction department, the incorporation of a second window between the order box and drive-thru window was designed at an estimated remodeling cost of $8,500 per restaurant.

In order to test these ideas, both the single and double window drive-thru's were simulated. One of the immediate hurdles was that while Burger King maintained detailed financial data and records, there were very little operational data available. Everything had to be collected and processed from scratch.

Burger King Corporate Headquarters had eight designated Research and Development restaurants located in its vicinity. These were the locations where data were collected. These data included but were not limited to interarrival times for customers at the drive-thru, time to make change, and time at the drive-thru window.

The results of the simulation corroborated the anticipated results. It appeared that for a per-restaurant cost of $8,500, Burger King could increase its drive-thru sales capacity by 15%.

3. Presentation of the Results

To move an idea from chapter to becoming a Burger King Standard required that approval had to be gotten to implement it into one test restaurant and, if successful there, that it be implemented into more test restaurants. From there, if the idea proved to be successful, it had to be tested in at least one test market in both company and franchise stores. Of particular note is that in order to implement an idea such as the double window drive-thru, the physical building requires remodeling, and crew members and restaurant managers require training which, in turn, requires that training manuals as well as trainers be developed. In other words, it is a major undertaking.

Because data collected at each of the above stages supported the hypothesis that the double window drive-thru increased drive-thru sales capacity, the concept was included in the decision to define the new Burger King restaurant standards. This was arguably the most important decision that management had to make since it was a binding agreement for all new franchisees as well as a requirement for renewal of existing franchise agreements. The discussion of the new proposed standards was the subject of a multiday officers meeting (all individuals holding the title of Vice Presidents (VPs) and above).

During the meeting, each individual change to the standards was to be presented and discussed by its "champion." I was the champion for the double window and presented the concept, its rationale, and the results obtained in the test market. The presentation followed the steps presented earlier in this chapter. Two key visuals (Figures 5.2 and 5.3) were used for stage 4 of the presentation: drive the audience to action. The first, recreated in Figure 5.2, summarized *why* the double window created benefits. While a traditional presentation might have elaborated on assumptions, methodology, and caveats, these were simply not germane to top management. You were considered the expert, and you were expected to have made sure that your staff did what needed to be done professionally and competently. Management at this level wanted to know where the beef was, so to speak.

This slide was designed to create understanding. Without understanding, no audience will be driven to action, hence its criticality. However, once understanding has been achieved, an equally critical

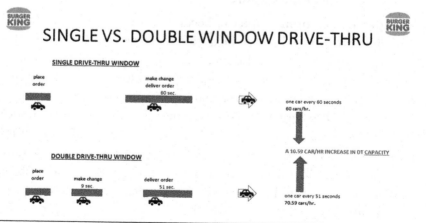

Figure 5.2 Why the double window drive-thru works.

IMPACT OF DOUBLE WINDOW DRIVE-THRU

INCREMENTAL CAPACITY	10.59 CARS/HR	10.59 CARS/HR	10.59 CARS/HR
HOURS/DAY AT CAPACITY	1.00	2.00	3.00
AVERAGE CHECK SIZE	$ 2.50	$ 2.50	$ 2.50
INCREMENTAL DAILY SALES	$ 26.48	$ 52.95	$ 79.43
INCREMENTAL AVERAGE ANNUAL SALES	$ 9,531.00	$ 19,062.00	$ 28,593.00
COST OF ADDING DOUBLE WINDOW	$ 8,500.00	$ 8,500.00	$ 8,500.00

Figure 5.3   Impact of double window drive-thru by time-at-capacity.

step is to present the impact of the decision in business terms that can equally be understood by top management and the smallest franchisee. Figure 5.3 is a recreation of the visual that was used to drive home the impact of the double window drive-thru. A key point that was the additional sales would only be realized during times when the drive-thru was full (e.g., at capacity). Hence, the impact is given as a function of the total time during the day when the drive-thru is at capacity. A large proportion of restaurants did achieve drive-thru capacity at least one hour during the day considering that, during the burger wars, 50% of total sales for the average restaurant came from the drive-thru. Another key point is that during the Burger King wars, day-to-day sales (today's sales compared to the same period last year) was the paramount performance criterion used by top management and, consequently, by everyone in the system. I was so amazed by that which I asked the CEO at the time whether profits mattered. The equally amazing response I received was that if your sales increased, profits would follow. This focus on the top line is why Figure 5.3 contains only sales data. Every corporate office and franchisee focused on that number and was able to internalize its meaning in terms of profitability. Furthermore, since franchisees had their own investment criteria, no specific information was given in terms of payback time, etc.

After the presentation, it appeared that the audience—all officers of the company—were driven to action, meaning that they would include the double window as a standard for all Burger King restaurants. During the ensuing of the presentation, the Senior Vice President of Operations, who reported directly to the CEO, asked whether the same results, as obtained with the double window drive-thru, could be obtained if all restaurants were operating according to standards.

This meant, in terms of the earlier example, whether single window drive-thru's could make change and deliver the order to the customer in 45 seconds. Our simulation models, when populated with the standards, did show that when operated according to standards, the 45-second window time could be met. However, our biannual QSCV survey showed that nowhere was there a Burger King restaurant that was operated according to standards (The QSCV study consisted of anonymous data collection at a representative sample of Burger King and nearest competitor restaurants across both national and international markets). My answer to his question was "yes, but we have yet to find a restaurant that operates according to standards. The double window drive-thru will bring existing restaurants as they are operated now to a 50-second window time."

After the meeting, the Senior VP took me aside in private and told me "I'll be g-dammed if I am going to fix an operational problem with a capital expenditure. I want you to go into tomorrow's meeting and argue against the double window drive-thru." After having had successful results (by my account) time and time over and having presented them to all officers, I was taken aback at the Senior VP's reaction and in a serious dilemma on how to maintain my integrity while not risking the Senior VP's wrath. I managed to do both by pointing out during the next day's discussions that the double window drive-thru increased sales *capacity* and that the benefits would only be accrued by those restaurants operating at capacity. I also pointed out that all restaurants could improve their sales capacity by operating according to standards. My arguments apparently worked because I did not get fired and the double window drive-thru was adopted as a voluntary option in the new Burger King restaurant standards.

4. Lessons Learned

Institutions such as the Institute for Operations Research and Management Sciences (INFORMS) have referred to Operations Research and Management Science as the Science of Better (www.scienceofbetter.org/). First, as students, we are brainwashed to think that just because it is better, it will be adopted and embraced by the organization. As teachers, we continue to perpetuate this mind-set. But, we fail to realize that what is made better was somebody's responsibility in the first place and that any suggestion of improvement in their domain has the risk of being viewed as a failure to perform on their part. This was the case with the Senior VP related above. He realized that the double window drive-thru would impose a capital expenditure on the corporation which could be avoided if he could enforce Burger Kings operations standards across the corporation. This, of course, begs the question of why they were not enforced to begin with. The above can be paraphrased into what I will dub as Swart's First Law of Improvement, namely "Any improvement suggestion elicits an equal, but opposite reaction from those responsible for making the improvement in the first place."

Understanding Swart's First Law of Improvement is critical in communicating with top management and driving them to the desired action. It points out the need for knowing your audience and, more importantly, understanding the impact of what is being communicated on those who are responsible for the area that is targeted by the improvement. From the communicator's perspective, the old adage of "being forewarned is being forearmed" is true.

Case Study 2—Bennigan's

1. What Was the Problem?

Bennigan's, like Burger King, was part of Pillsbury Corporation's Restaurant Group. They considered themselves to be a drinking-oriented eating establishment and thus quite different from Burger King. Burger King's accomplishments with Operations Research had received substantial publicity including a segment on Good Morning, America, and a write-up in Restaurant & Institutions magazine.

Intrigued by the publicity, Bennigan's management approached Burger King's management about using their expertise to assess the benefits of Operations Research in their organization. By this time, Burger King had expanded the original Operations Research Department to include Industrial Engineering (IE). This combined IE/OR group was asked to provide the requested assistance to Bennigan's.

From a personal professional perspective, I relished the opportunity to undertake this project. Discussions with Bennigan's had implied that if the results indicated potential benefits for Bennigan's to use Operations Research, then they would consider funding positions in Burger King's IE/OR department to support Bennigan's projects.

This, I hoped, would lay the pathway to the creation of a consolidated IE/OR group for Pillsbury's Restaurant Group.

2. What Analyses Were Performed?

Discussions with Bennigan's management revealed that their restaurants were at capacity during the peak hours which included lunch and the happy hour/dinner time frame. Furthermore, capacity was determined by table turns. This meant that if the time between when the customer was seated and when the customer departed could be reduced without negatively impacting the customer's experience, the probability that the table could be reused was increased with an accompanying increase in revenues.

To uncover opportunities to improve table turns, a comprehensive work sampling, also known as activity sampling, study was designed. Activity sampling is tantamount to having observers surveying all areas of the restaurant and recording the state of each employee and each resource. Providing that enough observations are taking to make the results statistically significant, the analysis of the results indicate the percentage of time that each employee was engaged in each activity and the percentage of time that each resource was used.

The study was conducted in a group of representative restaurants, and the results revealed a number of opportunities to improve table turns that had the potential of increasing overall restaurant sales capacity in excess of $7 million/year.

3. Presentation of the Results

The results exceeded the expectations of Bennigan's management, and a presentation of the results was scheduled during Bennigan's next officer's meeting. The format of the presentation followed the steps outlined earlier in this chapter. The audience, consisting of every VP and higher, was totally engaged with the presentation and, at the conclusion, appeared to be totally driven to implement the ideas that were presented. My contact at Bennigan's, the CEO's right-hand man, affirmed the impact of my presentations and indicated that I would soon hear regarding the next steps to be taken.

To make a long story short, after several attempts at reaching my contact, I finally caught up with him. I expressed my surprise at not having heard anymore after what I had perceived as an extraordinarily successful presentation. He responded that yes, indeed, my presentation had been extraordinarily successful. Virtually, the entire top management team agreed with the opportunities that had been uncovered as a result of our study. They all agreed that while they had suspected that things were not as good as they could be, now that these opportunities had been so vividly described, they felt that it was their duty, as officers of the corporation, to find ways to take advantage of the opportunities that had been revealed.

4. Lessons Learned

The presentation revealed numerous opportunities for improvement. The audience basically said "thank you very much, we'll take care of it ourselves." While this was an unanticipated reaction to the presentation, it also gave rise to what we may refer to as Swart's Second Law of Improvement which may be stated as "Any management group will continue to perform as usual unless it is compelled to change by the action of an external force."

In order to realize the opportunities for improvement revealed to Bennigan's management, some of the same tools that were used at Burger King (e.g., simulation and optimization) would be required. By opting to pursue these achievements on their own, Bennigan's management had committed to use their existing knowledge (or lack thereof) to fix the problems that their know-how created. This leads to the following corollary to Swart's Second Law of Improvement: "When compelled to change, management will continue to perform in the manner that created the need for change in the first place."

Both Swart's Second Law and its corollary are crucial in the communication of analytics to top management. While the major goal of communications is to drive the audience to action, it must be the correct action. In Bennigan's case, the audience was driven to take action. However, the desired action was not in having them try to achieve the results that analytics would have given them. This can be avoided by focusing on step 3 in Figure 5.1 and emphasizing two elements: (1) the opportunities that are being presented to you are not there because of you (management) not doing your job. They were found using analytics, a new toolset to which you had no access, and (2) achieving these opportunities requires the use of this toolset by individuals who have been trained to use the tools.

By prefacing step 4 with step 3 as described above, the groundwork has been laid to drive the audience to the desired action which is to adopt the recommendations presented as executed by trained analytics professionals.

Conclusions

Communicating analytics results to top management is not what I learned in school. Before going to Burger King, I had achieved the rank of full professor in academia by communicating my research results to my peers. I had been steeped in viewing effective communications as defining what is via thorough literature searches, stating what I intended to contribute to that body of knowledge as precise research hypotheses, testing those hypotheses according to the scientific method, and then presenting my results. Doing so successfully meant a publication, the undisputed proof of effective communication in academia.

That process became "academic" as I conditioned myself to my new environment. Management could care less about literature searches and did not have the time or inclination for written reports. A half-hour presentation was the typical time allotted to tell them what you had to say. You were being paid for your expertise, which had to be defended by answering whatever questions arose. Failure to do so was career limiting.

As I progressed from a BS in Industrial Engineering to a PhD in Operations Research, optimality was the Holy Grail. It was tacitly assumed that an optimal solution was going to be embraced and implemented based on its own merits. I had never been told nor had it occurred to me that being engaged in the science of better, as Operations Research is referred to, meant better than someone else had been able to achieve. And, someone else was not going to willingly embrace or support the implementation of whatever it is that I was recommending.

Once a presentation had been made to top management and the recommendation had been accepted, then the person responsible for the area—often also a member of top management—would be told to "make it happen." The person who received the command to "make it happen" is the de facto person for implementing the recommendations.

Unless top management has specifically indicated that you are part of the implementation team (a highly unlikely event if you are a consultant), then the implementer will generally continue to perform in the manner that created the need for change in the first place.

These thoughts provide insights into how analytics results may be received by at least some members of top management. To be forewarned is to be forearmed. Thus, these ideas should be an integral part of how to analyze the anticipated audience and how to organize the presentation so that the results will drive top management to take the desired actions.

6

PROJECT PRESENTATIONS IN THE ARMED FORCES

LYNETTE M. B. ARNHART

U.S. Army, Retired

Contents

> Military Briefings are not known for being clear, concise, or compelling; in fact, far from it. They are typically painful PowerPoint presentations filled with bullets and bad charts, low resolution images, and poorly formatted graphs.
>
> **Joseph McCormack, [brief], pg. 119**

In the armed forces, limited time necessitates that leaders are informed of the bottom line at the beginning of a briefing. This principle is captured in the acronym BLUF (bottom line up front). Stating the BLUF enables the decision-maker to discern the objective of the presentation being made—information, recommendation, decision, or request for guidance—and enables him/her to shape their thoughts and questions accordingly.

Unfortunately, unclear and painful briefings are indeed more common than I would like to admit. As an analyst, I would like to think that all our work is clear, concise, and of immediate and compelling value to our decision-makers. However, having created my share of slides and having reviewed and shaped far more, I cannot allow personal vanity to stand in the way of recognizing the truth in Joe McCormack's words.

In my experience, the armed forces remain reliant on PowerPoint to communicate. PowerPoint is a crutch, including in communicating analysis. That said, a crutch was designed to aid in capability, not to hamper. In this chapter, I describe some of the many types of analysis conducted in the military and how each of those analyses is used. I will also describe challenges associated with communicating analysis including time, depth of analysis, complexity of the problem, and vagaries of multiple audiences. I will wrap up the chapter with some techniques for presenting analysis using PowerPoint as a communication tool.

Overall, an immense opportunity exists to communicate analysis better. Good communication skills are directly related to success. The late Dr. Robert E.D. Woolsey once told me that an 80% solution implemented was better than a 100% solution on the shelf. Over the years, I have returned to this thought many times. Eventually, the quote led me to realize that someone who can communicate a partial solution well will almost always succeed over someone with a complete solution who cannot communicate what they have. In this chapter, I focus on lessons I have learned about presenting analyses in the military. These techniques are not peculiar to the armed forces but rather apply broadly across the many types of analysis in the private, nonprofit, and other government sectors as well.

Common Types of Analysis in the Armed Forces

The types of analysis used in the armed forces are many. Some of them are as follows: requirement analysis, analysis of alternatives, cost analysis, personnel analysis, budgetary analysis (planning, programming, budgeting, and execution (PPBE)), combat modeling, simulation, logistics and medical analysis, and ad hoc analysis. While there are likely many more types of analysis conducted, these represent some of the most common.

Requirement Analysis is used to analyze future concepts and requirements for operations and organizational structure in a joint context using doctrine, organization, training, materiel, leadership and education, personnel, and facilities solutions.

Analyses of Alternatives (*AoA*) is the follow-on analysis when a requirement analysis determines that a materiel solution (computer technologies, military equipment, and/or weapon systems) exceeding a set dollar value is needed. An AoA is a complex and multifaceted analysis that discerns costs and trade-offs between multiple potential and existing solutions. This analysis is required before certain milestone reviews that allow the procurement to proceed. An AoA analyzes the advantages and disadvantages of the capability under consideration, models and analyzes system and operational effectiveness, and compares the proposed acquisition to sufficiently feasible alternatives, fully burdened life cycle costs, technology risks, energy consumption and training requirements.[1] System effectiveness generally compares the performance of the proposed piece of equipment to the existing equipment and to alternative possible systems. Operational effectiveness models the effectiveness of the proposed acquisition in military operations and compares and contrasts the impact the system has on the operation with the alternatives. An example of this is the procurement of a fighter jet, battle tank, ship, or truck when the current systems are outdated, ineffective, or worn out. An AoA may take several months to over a year and is generally conducted by a team of teams with specialized expertise with participation from multiple agencies.

The U.S. Army Cost and Economic Analysis Center defines *cost and economic (C&E) analysis* as the process of analyzing and estimating incremental and total resources required to support past, present, and future forces, units, systems, functions, and equipment and is to be used as a tool to help inform resource requirements and decisions.[2] Again, there are many subtypes of analysis that fall within this category. C&E analysis includes the development of independent cost estimates for weapons and information systems; and analysis of organizational structure, weapon systems operations and support, personnel, and installation (army forts, air force bases, etc.) costs. C&E also includes production of service (army, air force, navy, or marine corps) cost positions. Cost positions are used in the development of budgets and planning and programming for the future. Programming is the development of 7-year forecasts of the monies required to operate each

[1] DOD Instruction 5000.2 dated 8 December 2008, page 58.
[2] www.asafm.army.mil/offices/linksdocuments.aspx?OfficeCode=1400 and www.asafm. army.mil/Documents/OfficeDocuments/CostEconomics/Guidances//cam.pdf

service. The final document produced is called the Program Objective Memorandum (POM) and is provided to the President and Congress.

Personnel Analysis includes all types of analysis used to help manage the military manpower requirements of the services. To manage personnel, each service must acquire (recruit), train, assign, employ, retain, promote, deploy, and transition soldiers, sailors, airmen, and marines. This means that personnel analysis includes analysis on topics including strength forecasting, recruiting, losses, promotion, assignments, and much more. As an example, the armed forces require forecasting to estimate how many people will leave the service over the 7-year planning and programming horizon to inform analysis that in turn determines what recruiting and training requirements will be needed to be in the future. Further, changes in policy must be analyzed to determine how it will affect the force. Each of these affects the POM planning to ensure that budgets for military manpower are sufficient to cover all personnel costs.

Budgetary Analysis relates to PPBE of the monies needed to operate a service. Each service in the armed forces is required to submit a budget for the next fiscal year, an estimate covering the current budget year and the following year and a POM covering the next 7 years. Since these documents cover all aspects of a service's operations, facilities, maintenance, research and development, and pay and personnel costs, there are large numbers of military analysts with expertise in these areas. This type analysis is often divided up into subareas to enable the development of expertise. Further, each service does this analysis differently as service culture plays into each service's priorities and strategy which impacts the resulting areas of analytic emphasis yet results in the submission of the same set of documents from each service.

Combat Modeling, Simulation, and Analysis are used in several ways. All services model military organizational structure (organizational or force structure refers to units such as companies, battalions, brigades, fleets, squadrons, wings, and numbered air forces and armies) to gain insight into how certain employments and campaigns might perform against certain assumptions about a given adversary. This type analysis generally is based on the simulations of operations. The services have a number of models they use to conduct modeling and simulation—some are commercial, but many are proprietary to the government. Such analysis allows insight into friction points, gaps, risks, problems with assumptions, timing, allocations, priorities, and resources.

Logistics and Medical Analyses relate to the provision of resources and healthcare to the forces. Much of the logistics analysis focuses on ensuring that our forces have what they need when they need it, similar to just-in-time supply. It also includes the transportation of personnel and equipment to faraway places. A couple of interesting historical notes demonstrate the immensity of the scale of logistics analysis to support military operations. The Red Ball Express (the historic Allied Logistics formation to supply the European Front in World War II) was composed of 5,958 trucks and trailers that delivered over 824 million pounds of supplies from August 25th to November 16th, 1944.[3] The convoys themselves consumed up to 300,000 gallons of fuel daily.[4] Of further note, the comparatively small North African campaign required 10 million gallons of gasoline. Medical analysis relates to the planning and programming needed for the provision of healthcare (the armed forces also engage in significant amounts of medical research and analysis related to medical research). This analysis is generally based on factors such as the size and health of the current and future force, current operations, and plans for potential future operations.

Ad hoc Analysis, sometimes referred to as "back of the envelope" analysis, is usually short notice with limited time to execute. While this is considered an analytic sin in some analytic communities, I have included it here because there are many times when military leadership needs answers within minutes to hours and often in these cases, quick, assumption-based analysis is better than no analysis. This works best when the analysis comes from subject matter experts who are very familiar with the subjects in question. With any ad hoc analysis, it is essential to ensure that the receiving audience understands the caveats and limitations to the provided analysis. An example from personal experience, our high-level military commander while testifying before Congress had his support team call back to the analytic team in theater (deployed) for an estimate to answer a question he'd been asked. Our team had 20 minutes to do the analysis and return an estimate.

---

[3] www.transchool.lee.army.mil/museum/transportation%20museum/redballintro.htm
[4] www.skylighters.org/redball/

Audience, Time, and Complexity Considerations

The composition and expectations of the audience are the most important factors to keep in mind when communicating the results of analyses. The type of analysis is also a consideration. Each type of analysis generally comes with its own associated time frame. AoAs may take a year or more, and much personnel analysis occurs on a monthly or annual basis. Some studies take multiple years. Ad hoc analysis may take minutes to hours. Time frames are generally reflective of the complexity of the study or analysis, and each brings its own unique challenges. This section will provide some thoughts on each of these, beginning with the recognition that communication starts before the analysis begins.

The Audience

First, it is important to understand the problem to be analyzed. Within the armed forces, analytic teams often exist to assist other organizations or staff sections with problems they may encounter. When the problem owner brings a "problem" to the study team, it makes the problem owner the study sponsor. It is very important for the study team and the study sponsor to develop good communication regarding the problem to be addressed.

To start this process, the study team should restate the study request back to the sponsor as very often, the "problem" nominated for study is a symptom of some other unrecognized or unaddressed problem. An analogy I often use for new analysts is the difference treating the symptoms of a cold (congestion/sore throat) versus treating the cause (virus). A historical example of this occurred in the late 1990s in the army. The force was severely short Sergeants (grade E-5). The analysis team was asked to adjust promotion targets to promote a greater number of Specialists (grade E-4) to mitigate the shortage. The analysis team found that the problem was far deeper than just an insufficient number over promotions over the past year. Analysis and collaboration with the policy, financial resources, human resources, and training teams showed that the lack of promotions stemmed from insufficient numbers of eligible soldiers. The root causes were policies, both fiscal and training, that had been put in place 3–5 years earlier. The real problem was not a shortage of Sergeants; it was poorly understood and managed policy that manifested as a shortage of Sergeants. This example also shows the need to do good, in-depth analysis before enacting major policy changes, so the impacts are understood, communicated, and expected.

Therefore, to best help, the analyst team should ask what the study sponsor is seeking to achieve. This understanding helps ensure that the team is solving the right problem. It also helps the team see the issue from the sponsor's view which enables them better determine which potential solutions might be acceptable. As the team proceeds, they may discover and underlying problem. If so, it is essential to communicate that back to the study sponsor and convey how analyzing the seemingly unrelated or unimportant problem relates to the effect the sponsor is trying to achieve.

The second point is managing the sponsor's expectations. This means keeping the sponsor informed on the methodology, how long the study is expected to take, how progress is going, what impediments have occurred, and how the impediments have impacted the time line and updates on in-progress work. This continuous communication with the sponsor allows for changes that the sponsor may desire during the course of the analysis and prevents the analysts from solving the wrong problem. Solving the wrong problem in the military has lasting effects in two ways. The primary effect is making the analysts within the organization irrelevant. This effect can easily become self-perpetuating as many organizations are composed of combinations of military persons who rotate every 2–3 years and long-term civilian employees. The secondary effect occurs as military leaders move on to other assignments and carry their negative impressions with them, potentially eroding their confidence in the work of analysts at their new assignments.

In the past, another communication issue occurred when the immediate sponsor understood, accepted, and agreed with the analysis but failed to build the same understanding in their higher-level leadership. When that leader was later briefed by the analyst at the higher level and without the presence and participation of the study sponsor, the higher-level leader inferred that the analytic team was incompetent and argued that the team had engaged in fanciful modeling that was not representative of the organization's plan. The leader was not prepared by his staff to hear results counter to his intuition and understanding of the plan. Further, the leader did not understand that his team had provided the data for the modeling to the analytic team. That briefing was an unmitigated disaster. But it was also a learning point. The team learned that regular engagement and updates between the analysts and the higher-level leadership build good will, trust, and a greater willingness to accept study findings or results that the sponsor would otherwise prefer not to hear. This third point has become a best practice for many of our analytic teams as it allows leadership to see what they are analyzing, provide input on assumptions, and engage in questions and shape the analysis before it is final.

Part of the ambush above relates to a failure on the part of the analyst to know the audience. In that particular case, the work had been through a peer-review process, but the team did not understand that the higher-level leaders who were not colocated were unaware of the work. This experience helped build the understanding that to effectively communicate analysis, the analyst needs to have the best understanding possible of the audience. I recommend two solutions for this problem. First, develop an organizational procedure that formally captures the scope of the project and documents it in a client/analyst study plan agreement. While such agreements are used broadly in some offices, they can benefit nearly all analysis teams by setting the terms, time lines, and expectations for studies. This is especially useful when there are multiple analytic agencies collaborating on complex projects or when working with partner agencies to augment organic expertise. Such agreements generally are approved by higher-level leaders and as such ensure their awareness before significant investment in analysis begins.

Another solution comes from common military practice. Before and during operations, we conduct reconnaissance to develop a better understanding of the adversary and what they may do. For briefings, the same mind-set is valuable. Learning who will be in the audience, what their views on the subject matter are, how they view analysis, and what knowledge they have of the ongoing analysis helps the presenter prepare for the briefing. Knowing the audience also assists the analyst prepare for potential countering views or positions and provides insight for the time allowed for the brief. Further, check to see if any of key individuals in the audience is colorblind. If so, structure the brief accordingly.

Each individual arrives to a presentation with their own bias and agenda. Some audience members will have objectives that conflict with those of the presenter or other audience members. Conflicts between audience members may put the analyst in the position of having to manage conflict and manage multiple objectives. Simple expectation of this occurrence is seldom sufficient to prepare an analyst to be successful during a presentation. Gaining insight to the audience, their perspectives, concerns, biases, and objectives takes time but pays off in the brief.

There was an army analyst who was masterful in his handling of audiences. After learning the primary objections to the work he would be presenting, he often wove it into his brief and allowed the dissenter to be his transition. His talk track went something like this, "I am glad you asked that question. I will show you why that is not the case on the next slide." This analyst very effectively diffused arguments by weaving them into his presentation and addressing the equities of the parties involved, thereby turning what could have been contentious into collegial discussion.

For those of aspire to this level of achievement, pre-briefing key decision-makers or influencers can help analysts gain insight into concerns they may have about the analysis and helps the presenter prepare the final brief. Whenever possible, a briefer should take a notetaker and a watcher along. The notetaker captures the context of the conversation and frees the briefer to focus on the presentation. The watcher's job is to watch the faces of those receiving the presentation and note where potential problems may lie based on nonverbal gestures and expressions.

Once the insight is gained, these team members should help the presenter prepare by taking roles for rehearsals. Rehearsal is key to success for every presentation. No matter how many times an analyst rehearses in his/her head, the words that will come out differently went they are actually said. The second piece is that rehearsal allows the presenter to become so familiar with the material that they can easily adapt as presentations rarely go as planned. The final comment on audience is that a dress rehearsal in the actual room scheduled using the actual equipment and slides lends itself well to reducing anxiety and potential information technology issues associated with briefing. The next part of audience management revolves around ensuring that the presentation is right for the audience.

TIPS AND BEST PRACTICES FOR DEALING WITH AUDIENCES

- Restate the problem back to the study sponsor.
  - Ask the objectives of the study sponsor; understand what they are seeking to achieve/accomplish and why.
- Clarify and get agreement on scope, time line, and deliverables.
  - Have them sign off on key requirements when possible.
- Schedule and conduct regular in-progress reviews to keep leaders informed.
  - Discover problems early and generate support/acceptance of work.
  - Manage study sponsor expectations by keeping them informed
- Pre-brief key higher-level decision-makers and influencers.
  - Be flexible to changes in the study's direction or scope.
  - Use a notetaker and a watcher for in-progress reviews and pre-briefs.
- Anticipate questions/countering views, and work them into the presentation.
  - Expect some people will continue to disagree. Determine ahead of time who will be in the audience and what their equities, concerns, objectives, expectations, and issues are.
  - Learn if anyone in the audience is colorblind.
  - Rehearse, rehearse, rehearse. Build transitions into the rehearsal.

Complexity and Time

The complexity of an analysis impacts an analyst's ability to present it at a level appropriate to the audience. For instance, a briefing to another set of analysts such as a peer review will likely be more in depth and get into the details of the studies. For such a brief, the analyst should be prepared to discuss assumptions, why such assumptions were made and why they matter. The analyst will also have to discuss data sources, which entities verified or validated the data, which models were used and why, how the model works, and how the results were interpreted and formed into results or findings. The primary or lead analyst doing the work is generally best suited to do this type briefing. Analysts who have worked in the Pentagon have often found themselves doing this kind of briefing with auditors from the Government Accountability Office, professional staff members of Congressmen and Senators or for peer review. These type engagements can be collaborative and collegial or adversarial. Knowing which to expect helps in preparing the brief. However, this type of long brief preparation is clearly not suitable for shorter engagements as shown in this next example.

Several years ago, the army recently completed a study that took over 2 years. Despite the complexity, and the in-depth nature of the study, the analysts had only 30 minutes to communicate the study, findings, and recommendations to the Secretary of the Army. This is far from uncommon. Modeling and simulation studies generally take months to prepare, but the briefer often gets only 10–15 minutes to speak with leadership about the work. For analysts, the details are an important part of each analysis.

The immense amount of knowledge, data, and work that goes into an analysis often makes it difficult to condense an in-depth study down to the few key points to communicate the information that the leader needs to inform her/his decision. I have encountered two views regarding presenting analysis to higher-level leaders. The first is that the briefing should only present the study findings (no recommendations) to the study sponsor and leave the sponsor to determine the significance and follow on actions to "do something." The second is that findings should be accompanied by recommendations. In the military, my experience is that higher-level leaders want and expect recommendations as the logical follow-on to the work. Further, developing feasible, acceptable, and suitable recommendations forces the study team to work with the sponsor team. However, both views can be accommodated in the approach below if the analysts consider providing enough information to support decisions to be the "do something" part of the task.

Outlined below are two techniques that can help analysts condense and distill key information from studies into the fewest number of slides. The first concept comes from training conducted for military information support operations (MISO) practitioners. MISO officers are taught that their task is to influence someone to do something. Analysts can approach communicating analysis in the same manner.

When constrained to a limited amount of time to present and discuss analysis, it helps to limit the number of content slides to four to six; not more than ten and plan to talk for not more than 1–2 minutes per slide. This allows for accompanying discussion during the course of the presentation and afterward. To get down to this number of slides, analysts should consider and determine the minimum amount of information a decision-maker needs to understand the conduct and outcome of the study and the information needed to inform any follow-on decision.

Professional looking slides are not necessarily overly complex. The most powerful communication tools are so simple that their message is unavoidably clear. Therefore, avoid needless slide junk, commonly referred to as "bells and whistles," by using only the minimum amount of graphics, pictures, and text needed to convey the key ideas. If your slides become so text heavy that they look like a page from a book, it is an indicator that the analyst should be writing a paper rather than building slides. Likewise, limit the use of those techniques and colors that may be hard to see or read on screen. Sparing use adds greater emphasis when they are implemented. Reviewing the slides in the presentation medium, be it paper, portable document format (pdf), on screen, etc., helps ensure legibility. Further, only assumptions that are material to the subsequent decision need to be shared. All else can be put into the accompanying read-ahead package, final report, or supporting slides.

The second technique that is useful is ensuring that every main slide clearly communicates a single idea. This is called a "takeaway"; military analysts often prepare by asking themselves, "what is the takeaway from that slide" or "what is it that you want the decision-maker to take from this slide?" It can be useful to create a "bang box" or "tone box" displayed on a banner on the bottom or side of the slide to highlight the key information. Viewing each slide individually aids analysts in preparing a clear, powerful, and dynamic presentation.

Below is an example of a four-slide presentation that shows the overview of a study and outlines each phase.

Study Overview

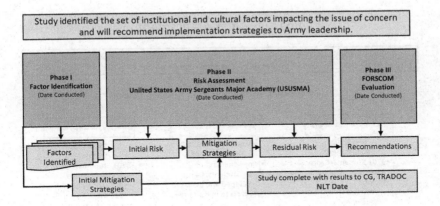

This slide shows the phases of the study, how they related to each other, inputs, and outputs of each phase and implies location for each bit. It helps the decision-maker understand the complexity of the study, indicates natural breaks for in-progress reviews, and sets the completion date. Green boxes represent activities, and gray boxes underneath represent products.

Phase I – Factor Identification

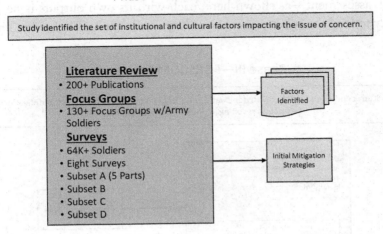

This slide expands on the first phase of the study. The top gray box outlines the purpose of the phase, and the green box details the activities and the magnitude of the work that was accomplished to produce the products. I have purposely kept these slides generic, but the analyst could build a bit more detail in the slide by detailing the factors and strategies identified.

Phase II – Risk Assessment

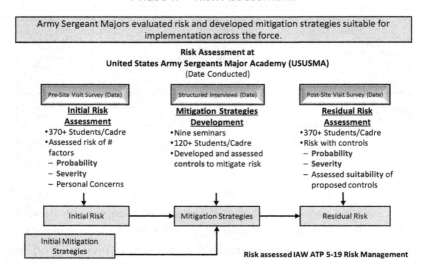

The risk assessment is clearly broken into three subordinate parts—initial risk assessment, mitigation strategies development, and residual risk assessment—as shown here; each with its own outputs is necessary to feed the subsequent action. Key words are highlighted in bold.

Phase III—FORSCOM Evaluation

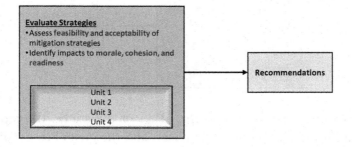

The final content slide of the set wraps up with the evaluation and presentation of the recommendations. Again, the analyst could add detail here on the specific recommendations.

For senior levels of leadership, the slides above may be combined into a single slide that encompasses the entire study as shown below. The power of the slide is that it can be used to any type briefing and moves easily between an initial brief, a peer review, an in-progress review or final out-brief at any level without needing modification. Any of the analysts involved with this study could talk to the study and go as in-depth or as high level (wave top) as needed. This slide[5] and similar overview or methodology slides help the sponsor and decision-makers understand the problem, plan, what will be/was done, and time line for the analysis. Moreover, the slide becomes a single slide "carry along" that the senior leader takes with him to use when he/she updates his/her boss or political leadership on progress. Developing and briefing slides like the one below that put together a complex study in a single slide helps the higher-level leaders set the analyst up for success, particularly if the leader is supervising multiple studies. Do keep in mind that the complexity of this slide means that for certain audiences, the multiple slide presentation may be a better option.

Study Overview

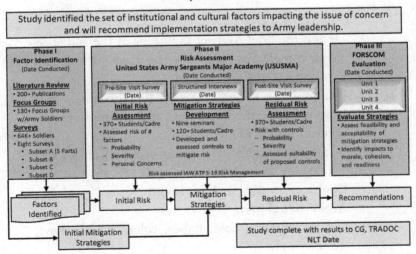

---

[5] Credit Training and Doctrine Command Analysis Center, Fort Leavenworth, Kansas, 2015.

Analysts must also be prepared to manage time during the briefing. This is a skill learned best through experience. Again, the rehearsal aspect is valuable. The presenter can have colleagues help them prepare by asking questions during the brief or by having side discussions that interrupt the presentation. The presenter needs to learn how to get the briefing back on track and in some cases how to summarize the key points when discussion has taken most of the briefing time.

Therefore, I generally recommend analysts use a ratio rule when planning a briefing. After learning the composition of the audience, the analyst estimates how much discussion is expected and plans accordingly. For a 30-minute decision briefing, I generally plan for 10 minutes of talking and 20 minutes of discussion. For information briefings, the talking time is generally equal to or slightly more than the discussion time. For presentations at professional forums, I try to ensure that I leave at least 10 minutes for questions and discussion. There is no ratio that meets everyone's needs—analysts need to develop their own based on their specific analyses, audiences and the audiences' expectations, and general past behavior.

TIPS AND BEST PRACTICES FOR DEALING WITH COMPLEXITY AND TIME

- Use BLUF.
  - Design the brief to meet the needs of the audience; adapt complexity based on the level of audience knowledge and time available.
  - Use pre-briefs to help shape the final brief.
  - Communicate the key points and assumptions relevant to the decision.
- Use the fewest number of slide necessary; four to six recommended for senior leadership.
  - Ensure every slide makes a point.
  - Avoid slide clutter.
  - Use underlining, bold, italics, color, 3D, shadows, etc. sparingly.
- Use backup slides and white papers in a read-ahead packet to add greater detail, if needed.

- Condensed overview slides can be immensely useful to senior leaders.
- Develop a plan to manage presentation time including time for discussion.
  - Plan only 1–2 minutes of speaking time per slide unless it is a condensed overview slide.
  - Keep an eye on the clock; ensure key points are communicated in the time allotted.

Other Examples of Successful Techniques and Slides

In the armed forces, slides walk. This means that slides are often transmitted, generally via email, far and wide. This can result in single slides being pulled from the deck and used by people who were neither part of their development nor previously briefed on the content. This dynamic makes it imperative that each slide can standalone in transmitting its idea or information. A classic example of this came in the mid-2000s when our office developed a strength history/forecast slide to brief the Deputy Chief of Staff of the Army for Personnel. The original complex chart was built in Excel but pasted into the deck as the picture. Despite being unable to alter the graph, it showed up in a subordinate command's presentation without reference to its origin or original purpose.

Analysts should also think about the graphics they use to communicate ideas. In the mid-2000s, the army was using a policy called Stop Loss to help ensure sufficient numbers of soldiers were available to fill deployed and deploying units. Under this policy, soldiers were held on active duty past their contracted active obligation to ensure the units were not short the soldiers needed to accomplish their missions. The analysts struggled to communicate to army leadership how the numbers of these soldiers diminish over time, meaning that the numbers held over for short periods of time were much larger than those held over for longer periods of time. The analysts eventually likened it to melting and used an ice cube analogy. The graphic they created used cubes dimensioned by height, width, and color to show the effect and was very successful in communicating their analysis of the policy and its effects.

While I no longer have the original (and much more elegant) chart, the one below illustrates the concept. It shows 12 monthly snapshots of the (notional) magnitude of the soldiers affected by the Stop Loss policy, how long they have been affected, and increases or decreases over time.

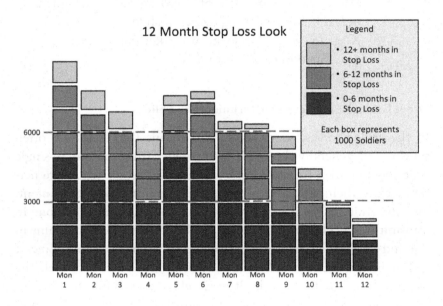

The wargaming slide below is another example of the presentation of complex ideas. The armed forces include wargaming in their analytic activities. Wargaming, a decision support tool using adversarial players and decisions, is generally used to test ideas and identify shortcomings, friction points, gaps, risks, and required resources. The chart shows the following:

1. The linear phases of wargaming.
2. The proportion of responsibility for each phase shared by the sponsor or the Command's Wargaming Cell.
3. Activities involved in each phase.
4. Example products from each phase.

As a result, this slide can be used as a methodology or overview slide for any presentation on wargaming. It can also be used as a template for wargame development irrespective of the plan or decision under examination.

Wargaming Activities by Phase

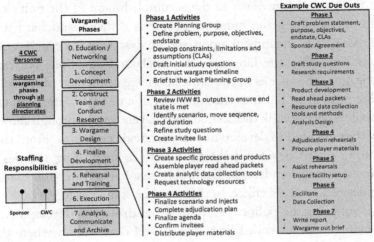

Another effective technique to help decision-makers understand modeling and simulation of operational plans is using embedded video that captures the simulation playback. Using 30–60 seconds of playback enables leaders to "see" their plan in action and helps them think through questions and concerns regarding key timings. Showing leaders their plan in action, per se, helps analysts assure leaders that the modeling is accurate and representative of the plan. It also helps decision-makers accept findings that are unexpected. For example, in a recent briefing to their commander, analysts identified four potential, yet previously unidentified, decision points in the operation as a result of modeling and simulation. Prior progress reviews with the commander using modeling and simulation analyses had built a level of confidence in the analysis team that allowed him to both understand and accept these findings.

Animation can also be used in presentations, but should be used sparingly and only when it adds to the overall presentation of the analysis. This option is only available when presentations shown on a screen rather than briefed from paper copies. A recent example of this was an assessment of risk related to changing military force presence demonstrated by digitally moving icons for ships, planes, and tanks on a map. The result was that the finding of the analysis was immediately apparent to the decision-maker.

In general, good presentation requires putting analysis into the language of the audience. Far too often analysts are viewed as the "numbers" people. Leaders and other staff are busy doing important things that contribute to the accomplishment of the unit's tasks and understanding the numbers people is difficult and hard work. Therefore, it is incumbent upon the analyst to not be a "numbers person" speaking the language of models and equations but rather a colleague who speaks in the same language as the rest of the staff.

The next slide[6] outlines the construct of a military campaign plan. The colors clearly walk the desired political endstates through showing how they are used to develop military strategic endstates. From there, planning teams develop military objectives supporting the desired military strategic endstates. Each objective has at least one subordinate desired strategic effect. In this manner, it is clear to see how each effect supports the political endstates that drove the development of the plan. The color-coded lines, boxes, and circles show where there are overlapping objectives and effects that support multiple endstates. The gray box around all the military endstates shows that each military endstate supports Political Endstate #5. Further, the graphic is useful for subsequent presentations as it allows the analyst to assess and explain the impact of the achievement of any particular effect during or after the conduct of the campaign.

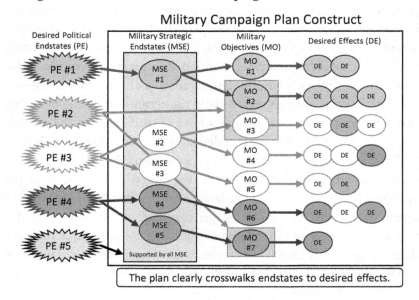

Military Campaign Plan Construct

The plan clearly crosswalks endstates to desired effects.

[6] Credit Coalition Strategy and Planning Group Assessment Team, MacDill Air Force Base, Florida, 2016.

When slides or graphics are sufficiently complex, three techniques can help. First, orient the listeners to the slide by walking them through the meaning of each part and the idea it is intended to convey. This is especially useful when presenting a series of slides which all use the same format, for instance, assessment of multiple objectives. The second technique is consistency—keep the meaning of colors and shapes consistent throughout the deck or, even better, across time. While the figures in this book are in a single color, the military consistently uses a red, amber, green rating scale. It is commonly accepted that anything red is bad, anything green is good, and amber (yellow) is somewhere in between. This means do not use green in a deck to talk to something that is generally considered bad (note: this is an example of preexisting audience bias).

The third technique is a printed handout (table drop) helps. Table drops are particularly useful when employed in coordination with complex slides or charts. The table drops used can provide specific examples from the plan being discussed and provide additional background information to which the listener may need to refer or may simply be an enlargement of a complex slide being displayed. Do note, however, that a table drop can also be a distractor from the main presentation. A useful technique for analysts presenting can be to manage the audience using language such as the following sets of words. To discuss the table drop, use "please direct your attention the paper in front of you" and then bring them back the main presentation with "if you would redirect your attention to the screen."

The ways to present analysis are unlimited. This is true within the armed forces as well. Be creative, think hard, try out possibilities, and determine what works best. Great slides don't generally happen in a single iteration or sitting but are often developed through hours of hard, collaborative work. I and another analyst once had to go see a very senior leader to explain a complex study we were working. We failed, but recognized that his preconception of the problem prevented him from understanding. We returned to our cubicles and brainstormed many ideas for other means to communicate our analysis. After designing an improved depiction, we returned. He sent us away. Again, feeling defeated, we once again returned to our whiteboard and brainstorming. Eventually, we developed a third graphic and sent the read-ahead packet to his office. When we arrived for the presentation, the leader explained our analysis to us! If a slide fails to communicate the analysis, try something new. Keep trying until finding an effective aid.

ADDITIONAL TIPS AND BEST PRACTICES

- Create versatile framework slides that can be used repeatedly.
- Use table drops to enlarge or add detail to presentation.
- Use animation thoughtfully—best used to express time, positioning, mapping, development, or other sequences.
- Use video playback of simulation or other analysis sparingly and only when it illustrates a key point.
- Communicate in the language of the audience.

Summary

Analysts must ensure they have mastery of the analysis tools and techniques and can clearly communicate the purpose of the briefing and the objective and findings of the analysis. The BLUF principle should be implemented as much as possible. The communication process begins before the project is launched and ensures that the analysts solve the right problem. It continues throughout the analytic process to build a better understanding of the vagaries of the audiences. It also assists the analysts in condensing complex analyses to provide decision-makers with the information they need in the language they understand. Thoughtful consideration, preparation, and rehearsal prepare the analyst with insightful graphics for dynamic presentation of a clear, concise, compelling brief of their analysis that enables informed decision-making.

7

INVENTORY MANAGEMENT

Customizing Presentations for Management Layers

ELYSE HALLSTROM

Intel Corporation

Contents

Inventory management is a critical component to any supply chain that directly affects the bottom line. If you do not have enough inventory on hand, you may miss potential revenue. If you have too much inventory on hand, you may increase your holding costs and scrap rates as products become obsolete and are replaced with newer ones. Inventory management and forecasting go hand in hand. Forecasting customer demand is a core function of supply chain management. The forecast lets the manufacturing department know what to make when and what components to procure ahead of time. Specific inventory targets by product are usually calculated based on forecasts so that as demand increases, inventory on hand has increased ahead of time and vice versa. One has to pay great attention to how the inventory targets are set up. The trade-off is between having too little inventory, which holds up production, and having too much, which increases cost.

Inventory Management at Intel

As an enterprise with $60B+ in revenue per year, Intel has a large material purchasing group with many subgroups focusing on materials for fabricating semiconductor devices, for packaging those devices for insertion into printed circuit boards, for acquiring spares for maintaining production equipment, and many more. We sell the microprocessors we make to our customers who then put them on a motherboard and into a computer, laptop, server, etc. to sell to their customers. Each of these commodities has their own version of the dynamic inventory-sizing problem. I was in the packaging commodity group. The packages we were concerned with here were the green substrates that the integrated circuits are put on and that have pins to allow it to plug in and connect to the motherboard. Each of these packages is unique to the product that we sell to our customers since some may have different processors on them, sometimes multiple processors, or different connections to meet our customer's needs. We procure these packages from outside vendors who make them to our specifications. Because they are all unique, we have to plan for hundreds of separate Intel products, each with its own distinct package, and each coming from multiple vendors. Since the products ranged from servers to desktops to laptops to tablets, the demand for each varied over time as did the demand for packages. These factors made the package inventory problem especially difficult.

As a data analyst in the packaging materials group, I was asked to create a recommendation on what the group's target range should be for inventory levels for the entire commodity. We maintained on-hand inventory to support our customer's orders but needed guidance on how much inventory we should keep in stock. Management used this range to assess the health of the commodity. "Health" was our ability to support our service-level agreements (SLAs) to our downstream partners in manufacturing, while keeping the inventory costs and scrap rate as low as possible. SLA defines the percentage of increased demand we can support within our substrate suppliers' lead time. We are able to support these "upsides" by maintaining appropriate levels of inventory on hand. The previous inventory range was established years prior and did not reflect the current economic conditions that we were experiencing. Management was afraid that if we were operating under old data assumptions that we would have a miss that could result in us holding up manufacturing by not having the right components on hand to finish our products, thereby potentially impacting revenue. There were already target inventory levels for each individual product that drove ordering based on desired safety stock and internal forecasts, but management wanted a range that was historically based for this high-level health indicator. The range needed to be flexible enough to support moderate demand fluctuations while supporting our SLAs within our organization.

The approach used calculated the target range based on several important factors. Months of historical weekly forecasts were compared against actuals to determine forecast error [(forecasted demand—actual demand)/actual demand], which gives us the percent we over or under called demand as compared to actual demand. Several service levels were calculated as well so that management could see how the SLA would impact the range and make further determinations on whether or not the SLA should be changed. Lead times by supplier were critical. These lead times varied by supplier and by product and varied over time as the supply chain improved, so the average and minimum values of the most recent supplier lead times were selected for each product. The lower bound of the range was from the minimum supplier lead-time calculation, and the upper bound was from the average supplier lead-time calculation.

There were four different levels of review that I had to go through in order to get to final accepted inventory range that the organization would enforce since this commodity was one of the most expensive. Being slightly off on our goal range would translate to millions of dollars, hence the high level of scrutiny. The challenge for me was therefore to make presentations to four different audiences with differing goals and technical knowledge.

Review 1: Presenting to My Manager

The first-level review was with my manager who wanted to take a first pass at making sure I had not missed anything, and to do a "gut check" that my range did not make him balk. We looked at the detailed calculation, breaking it down into its high-level components and then delving into the specifics of each part. Since this kind of calculation was my forte and not his, I intentionally tried to impress him by showing the underlying math, knowing that he would not fully track the calculations but wanting to demonstrate my capability as an analyst since I was new to his team. This may or may not work for you and you may not need it if you have already established credibility as an analyst, so think carefully before employing this approach by thinking of what you are trying to get out of the presentation and to whom you are presenting. After showing the full calculations, I focused on the details that I knew he cared about so that he would get what he needed out of the meeting. If I did not make him comfortable with my analysis and the results, I would not get the "go" decision I was looking for from him to move to the next level of review. He and I went through each of the variables that I used in the calculation, what level of granularity I used, and what sources I used for each. I also pointed him to the resources these calculations were based on so he knew my methodology was acceptable. This helped him gain comfort in the analysis at his level of understanding.

The most important part of this discussion was around how my proposal compared to the current range and why mine was different. Part of it was due to the difference in demand during the time these ranges were defined. The historical range was created based on previous operating ranges—it was not mathematically calculated. We argued that the only reason it worked for so long was because there was a downturn in the economy so we were seeing decreased demand. Our inventory levels were not being challenged with the normal upsides we would see. Since we were now in a more stable time period where demand was increasing and we were seeing the normal upsides we would expect, we were concerned that we would end up getting being unable to support our desired 98% SLA within lead time if we didn't recalculate the operating inventory range. The best way for me to explain this was with a trend of the data with delineations of the two periods. We also discussed how the changes in supplier lead times and the shift to a higher percentage of products with longer lead time had affected the calculation. Figure 7.1 shows the demand values for all products from 2012 to 2015.

We also reviewed a graphic of how the forecast error would affect the calculation. "Overcalling" means that the forecasted demand was

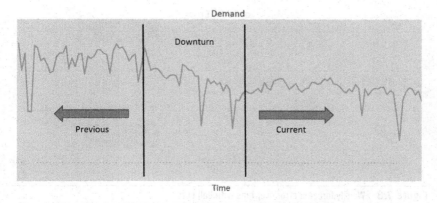

Figure 7.1   Weekly actual demand for all products from 2012 to 2015.

higher than the actual demand. This makes our forecast error positive, based on how we have defined forecast error; "undercalling" means that the forecasted demand was lower than the actual demand which makes our forecast error negative. When we were overcalling demand, the calculation would reduce the inventory range, and when we were undercalling demand, the calculation would increase the inventory range. Initially, I had created a simple trend just of the average forecast error over time as shown in Figure 7.2.

However, this required the audience to remember what each direction of forecast error meant. So then I broke out a historical trend and labeled it with each of these terms so that the conclusion would be immediately obvious to the reader instead of trusting that he would come to the same result. The arrows provide additional help in determining which direction we are calling out as shown in Figure 7.3.

Figure 7.2   Weekly forecast error over time.

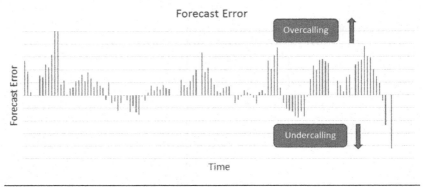

Figure 7.3   Weekly forecast error over time with call outs.

Another step further was to include the trend of the actuals over-laid with the forecast error so that it would add clarity to why the forecast error was positive or negative; the reader could see that we had overcalled the forecast when demand had dropped unexpectedly as shown in Figure 7.4.

There was a big change in forecasting methodology that could affect the forecast error and I knew there were going to be questions about—moving to an updated solver. The previous solver was not updated with additional business constraints and complexity over time, requiring a lot of manual manipulation required afterwards in order to reach a useable forecast. The updated solver comprehended these additional constraints and complexity, so manual manipulation was minimized. I knew management was going to ask how our forecast error had improved or changed since moving to the solver so in order to intercept any questions about this, I decided to call out that

Figure 7.4   Weekly forecast error and actual demand over time.

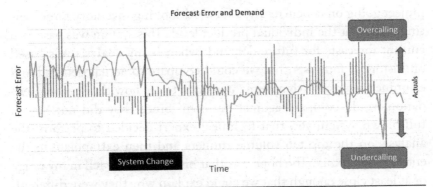

Figure 7.5 Weekly forecast error and actual demand over time with system change identified.

date in the trend data as shown in Figure 7.5. There wasn't any long-term discernable difference between the old system and the new system in terms of forecast error, but I wanted to call out the changeover date to head off any questions that might arise.

Overall, this graph of the forecast error with the demand overlaid really helped explain the result without having to show the actual calculations of the analysis. Anticipating potential questions around when the system change occurred in the trend helped to keep my conversation on track. It was also critical to call out details of over-calling and undercalling on the graphic instead of relying on the audience to come to the same conclusion as me. If I hadn't had these details captured, my review would have become bogged down in walking them through what each time period meant and leading them to the conclusion I wanted them to draw. By calling it out directly, they started where I wanted them to end and then got the support they needed by digging into the details necessary to back up that conclusion. This is a great tactic for presenting data to nontechnical people.

Review 2: Model Validation

Once my manager had approved my analysis and recommendation, it was on to the next review: a joint management and technical review. Since I was new to the group and had not built up a reputation to be a statistical analysis expert yet, management wanted to validate my result using a separate methodology as well. There was another project going on concurrently, a forward-looking inventory target setting system at the individual product level. This system was based on current forecasts for future demand, whereas my analysis was based on historical forecasts that we could compare to actual demand and examine the errors. Management wanted to see how their recommendation would compare against my own. Since they did not have a full-blown system yet, that technical expert decided to perform the analysis on the top ten volume runners and then extrapolate to the entire product mix. The hope was that his single value fell in my range or at least close enough that we could explain why they were different. There should have been some agreement between the analyses, but if they did not align, we knew what would be driving the differences. Since our methodologies were different, changes in the economy or demand could drive our results to differ. Also, changes in lead time over time could contribute to the difference since longer lead times drive increased inventory and shorter lead times enable reduced inventory in both of our methodologies. Additionally, since his method used the top ten volume runners and mine was based on the full historical product mix, there could be some small differences there if the top ten volume runners had different lead times, for instance. If our results had not aligned, we would have explained the differences to management as occurring from differences in our assumptions and mechanics. This was not necessary, however, since his number fell in my proposed range.

In this discussion, we only reviewed the variables considered in each analysis at a high level and the results. There were discussions around how our methodologies were different and may not align in the future as well. My manager was satisfied that this other analysis confirmed our recommendation, and it was on to the next hurdle.

Review 3: Technical Experts

The next review was a purely technical one. Upper level management, who would make the ultimate call on whether to go with my recommendation or not, had wanted to make sure that my analysis was blessed by known internal experts. They wanted to make sure that I had not missed any important variables in my calculation, used incorrect data sources, or used the wrong levels of granularity of the data.

This review was incredibly detailed. We went through the intricate details of the calculation, the assumptions I had made about where it was appropriate to aggregate and where fine granularity was needed. We looked at the data source for actuals to figure out how to align the data into the same time frame since the forecast numbers were offset from actuals by supplier lead time. We discussed how I handled the different supplier lead times by product, using the minimum and the average to define the range.

In this review, these internal experts did not care about the actual recommendation or the comparison to the current range—they only wanted to make sure my methodology was sound. There were alternatives I could have used instead, which they proposed and I explained why I had not done it that way. Either they agreed that I had chosen correctly or they accepted that the difference their alternatives would have created was negligible and did not merit reconfiguration of the analysis. One example of this was that I used the most current lead times by supplier by product. I did not find historical lead times to use in the analysis as this would add a lot more complexity to the model that we determined was not going to be significant enough to merit the extra time needed to create and run. Since lead times change over time, it would have added an extra dimension to the calculation to find the appropriate lead time to use for that specific forecasted week but without providing comparable calculation improvement. Since in the end we were aggregating to average supplier lead time, this complexity was dropped and the technical experts agreed that this was appropriate. In the end, I got their stamp of approval and was on the final round of review.

Review 4: Presenting to Senior Management

The final review was with upper level management to make the decision on what the range should be set at. The presentation was very tightly structured as a result of this focus on our parts to keep the discussion on track. This is where the bulk of the time was spent creating this presentation. If we did a poor job of ordering the information or got too detailed, we could spend the entire meeting focusing on the wrong things and end up without a new range to work from. Since upper level management's calendars are hard to get time on, it would mean additional weeks of operating under an out-of-date range that could put us at increased risk of meeting our service levels.

The first item discussed was what the expected outcome of the meeting would be: that we would settle upon a new inventory range to run the business on. This helped management know what we were looking for and helped focused them on the key issue. Without this expectation leveling at the beginning, it would have been very easy for management to get sidelined with details that could keep us from getting to the outcome we were hoping for. The other important component of this is that we had to make sure that the story in our presentation was consistent, succinct, and led our audience to reach the conclusions we wanted them to come to.

The next slide of the upper level management review was around the motivation for determining a new operating range. The previous range was not providing sufficient flexibility to cover the new upsides the team was anticipating. In addition, the previous range was not necessarily statistically based. It had been established around the current operating levels at the time and on what management thought was possible. Management had agreed in a previous meeting that this range should be investigated and adjusted based on statistical analysis. We took the time to remind them of this fact simply because sometimes management's memories are short since they are dealing with so many issues at a given moment. If they had been confused around why we were having this meeting, we wanted to clear it up immediately and give them the motivations so that they could remember the previous discussion and move forward on to reviewing the recommendations. This was obviously a very brief discussion, but critical to keeping moving towards our goal.

The next item on the agenda was to go through the recommendation and relevant technical details. We began with who had vetted the analysis in deep technical detail, their names, and organizations. This way management knew that it wasn't necessary for them to get into the weeds of the model themselves—there were trusted names who had approved it so they could just focus on the implications for the organization, not how we got there. We also mentioned that we had performed a separate analysis using different methodologies to compare our recommendation and their number was in our recommended range. This helped management know that the right people had reviewed the analysis and a separate calculation backed us up. This meant that they did not need to scrutinize our analysis; they could take what we were giving them as accurate.

We then moved on to the actual recommendation. In the review of the recommendation, we gave them a table of the different analysis we had completed with the different service levels. We had calculated the range based on our current service level of 98%, as well as of 95% and 99% for comparison. We knew they would not want to use the 95% or 99% service levels, but by having those numbers in the presentation, it helped make our recommended range seem more reasonable. Management considers a 95% SLA unsatisfactory, and 99% is too expensive, so those values justified our range additionally. In addition, we needed to give them one recommendation with a couple of alternatives. One way to do this would have been to simply include all the ranges calculated, as shown in Figure 7.6.

A better way to do this was with the recommended 98% SLA range highlighted with a box around to draw their eyes immediately to what we wanted them to see, as shown in Figure 7.7. From there, they could look at the alternative ranges and see that 98% was the range to go with to achieve our service level without inflating inventory unnecessarily. While adding a box in the table may seem like a trivial change, often small things like this make a large impact in improving audience understanding and response.

Service Level Agreement	Lead-time	Inventory Units
95%	Min	500
	Avg	700
98%	Min	613
	Avg	775
99%	Min	720
	Avg	830

Figure 7.6 Calculated inventory ranges based on SLA.

Service Level Agreement	Lead-time	Inventory Units
95%	Min	500
	Avg	700
98%	Min	613
	Avg	775
99%	Min	720
	Avg	830

Figure 7.7 Calculated inventory ranges based on SLA with highlight.

There were still ways to improve this table, however. Inventory range numbers have a "gut feel" associated with them, but it is not easily translatable into metrics management really cares about. In order to put it in terms of something that mattered to them, we worked with finance to determine what the cost of implementing this solution would be in dollars based on current product mixes as shown in Figure 7.8.

The recommendation was an increased range to work within, and we rightly anticipated that management would be wary of increasing our amount of inventory we were holding in dollars. In order to anticipate their questions around the current cost of inventory, we included those numbers below the graph in their separate table comparing the current inventory cost and the upper end inventory cost as shown in Figure 7.9. Since this was a ~20% increase in cost over the prior range, we also put in a mitigation plan to have inventory held in a different location to minimize holding cost to the company. We also had a net present cost calculation and explanation that it was due to the opportunity cost. It was important that in the previous slides, we had reminded management that the prior range was established during a down period and that we were now in a higher period of demand. This helped them anticipate that the recommendation was going to be an increase before they even saw the numbers.

The demand justification as well as the financial impact gave management everything they needed about the analysis in one slide. We had follow-on slides that we could go into if management was having

Service Level Agreement	Lead-time	Inventory Units	Dollars
95%	Min	500	$170
	Avg	700	$190
98%	Min	613	$185
	Avg	775	$210
99%	Min	720	$205
	Avg	830	$230

Figure 7.8   Calculated inventory ranges based on SLA with dollars.

Service Level Agreement	Methodology	Inventory Units	Dollars
98%	Current	570	$185
	Proposed	775	$210

Figure 7.9   Inventory cost between current and proposed methodologies.

a hard time with the increased range. These were from our finance partners showing the data to show that inventory is a cheap way to mitigate risk: the NPV of an individual unit was very low. If we ran into stock out situations, we would run the risk of losing market share as well as return customers, which would have a much larger impact on the long-term horizon. The net gain from our solution shown to management on another slide.

The math was only included to the credibility of our analysis. We only went over it at a high level, emphasizing that it was a good plan to increase inventory if management was concerned about the mitigating potential risk. No management would want to support increased risk if possible, so by phrasing it this way, it helped get them on our side that we needed to increase the range. There was an additional mitigation slide on how if we changed allocation strategies between our suppliers, we could find additional savings that could self-fund the increase in the range. Finally, we included historical scrap as a percentage of spending. This showed that scrap was in control and should continue to be so, based on our projections. In addition to being concerned about the increased cost of changing the inventory range, potential risk to increase scrap percentage is the other main concern we knew management would have so having these data readily available squashed any worries immediately.

Based on the analyses and presentations, we were able to get management to approve our recommended inventory range. The other critical part of our presentation is that we would revisit this recommendation twice a year. This helped management feel better that if demand changed, we would be responsive with changing the range instead of running at an inflated range compared to what was merited which would increase our scrap risk and costs. It was a full package: expected outcome, motivation, recommendation, supporting slides to handle anticipated concerns, and future cadence for review. We had addition details on the actual calculation in the appendix in case there were other questions, but we knew they were unlikely with the technical reviews already approved.

Summary

Each of the reviews had different objectives and dealt with different types of information. I therefore used different approaches for the

Audience Type	Presentation Tips
Manager (immediate supervisor)	• Include research if using new methodology • Cover the computation details if necessary • Compare your recommendation to current state in terms that matter to the manager
Technical Expert	• Show the details of your analysis • Explain sources of data and level of granularity use • Explain alternatives considered and why you did not choose them
Senior Management	• Start with your goal for the meeting and Include summaries of any previous conversations • Explain global parameters/changes influencing results • Compare your recommendation to current state in terms that matter (cost, scrap, SLA, etc.) • Anticipate questions and have back-up details on hand.

Figure 7.10   Presentation recommendations based on audience type.

presentations for each of the reviews. The first manager review meeting's goal was to make sure my recommendation passed the initial gut check and ensure there were not any glaring red flags in my analysis at a high level. The second meeting's goal was to compare my solution against a completely different methodology to validate my recommendation in a different way. The third meeting with the internal experts' goal was to vet the intricate details of the analyses, not the recommendation itself. The final upper level management meeting's goal was to approve the new recommendation. Since each of these meetings had different intents, there was a different roll-up of the data and flow in the presentations, shown in Figure 7.10. The technical reviews started with the low details and built to the recommendation, if the recommendation was included at all. The upper level management review only covered the motivation for the analysis and evidence that a technical team had reviewed it, before getting to the recommendation. The remaining supporting details came after the recommendation. It works best with upper level management to get them to end first and then build up the understanding afterwards. By anchoring their focus, we can walk them through the necessary details without getting them off track. The table in Figure 7.10 summarizes the key approaches to presenting to various audiences that I found useful.

8

EXECUTIVE COMMUNICATION IN PROCESS IMPROVEMENT

KEITH E. MILLER

Clayton State University

Contents

Introduction to Lean Six Sigma

Operational excellence is a key component of strategy execution, and programs designed to foster operational excellence play a both critical and dubious role in helping companies meet strategic objectives and fulfill missions. Just as Michael Porter's five forces model helped shape the modern approach to strategy planning, and Kaplan and Norton's Balanced Scorecard provided a critical link between organizational strategy and operations, the *Lean* and *Six Sigma* methodologies [hereafter referred to simply as *Lean Six Sigma* (LSS)] have provided structure and scientific rigor to process improvement. *Lean* traditionally focuses on eliminating process waste. *Six Sigma* focuses on variation reduction—treating deviations from a process target as a defect and eliminating or minimizing root causes. Since the early 2000s, these two methodologies have merged into a combined LSS process-improvement program.

LSS is a data-driven, project-based methodology organized into five logical phases: **Define** the problem and its impact on customers and business, **Measure** the baseline process performance, **Analyze** the process for root causes, **Improve** the process to eliminate or reduce root causes, and **Control** the improved process so it doesn't regress back to the old state, collectively referred to as the Define, Measure, Analyze, Improve, Control (DMAIC) approach. People play several roles in this methodology, each role with its own targeted training program. The roles include the following: **Champions** are process owners and business owners who control resources and have responsibility for operational performance. **Master Black Belts (MBBs)** are experienced practitioners who train others and manage process-improvement programs. **Black Belts (BBs)** (dedicated, usually focused full time on process-improvement) and **Green Belts (GBs)** (part-time, usually with a key role or "day job" in the organization) are practitioners who receive specialized training in process analysis, statistics, and quality management tools, and manage individual process-improvement projects.

LSS often involves intensive, potentially expensive (in time away from work and cost) training and requires cooperation of multiple stakeholders that necessarily have to continue managing their current workload. Therefore, the program carries certain expectations for return on investment. Six Sigma really took off in the 1990s when General Electric (GE) adopted the Motorola-developed methodology across the entire organization and touted billions of dollars in savings. Likewise, Lean is strongly associated with Toyota's marvelous transformation to world-class quality exemplar. With such amazing results, it's no wonder that a manager would have high expectations.

However (I am greatly simplifying here), communication breakdowns over time have contributed to additional barriers that a LSS practitioner must overcome to deliver a successful project outcome. It is easy to forget that GE spent billions up front, or Toyota spent decades to evolve the "Toyota Way." The three constraints of project management—cost, time, and scope—sometimes devolve into "no budget," "unrealistic time," and "scope creep." The first job of a LSS project leader is to communicate with a business leader to set realistic goals (re: SMART criteria of specific, measurable, achievable, relevant, and timely), and then to communicate progress and results in a manner to which that the leader can relate.

Data Availability, Level of Rigor, and Managing Expectations

It is very common for an LSS practitioner to be "parachuted into" a process with design specifications that do not include data generation for process-improvement purposes. It brings to mind a 1988 paper by DL Bourke with the apropos title, "Data, data everywhere, and not a thought to think." Out of nearly 450 LSS projects with which I have been directly involved, maybe three to five had relevant data, readily available for our needs. A project most often requires a data collection plan that frequently includes manual data collection resulting in a limited sample size.

Consider a BB project at a military installation focused on improving the outcomes of a strategic communication plan. No data existed to tell us how well messages were disseminated and understood. An organizational assessment indicated a weakness in effective communication. Measuring the penetration of messages from top leaders through all levels of the organization was not something they did. Rather it was assumed that when a colonel spoke, all parties received the message. Yet there was compelling anecdotal evidence to the contrary. So, the project first had to establish a viable measurement system, and then test it, before ultimately establishing appropriate channels for different types of messages. Sample messaging and manual monitoring demonstrated that no single communication channel effectively reached more than a third of the employees, and nearly a quarter of employees never received leadership messages. The project then focused on evaluating communication channels (manually, which was time consuming) and building the appropriate pathways for specific message types (e.g., issues regarding personnel, facility, and leadership). These time and effort burdens had to be factored into setting management's project expectations. Two keys to success, for this project and many like it, are given as follows: (1) incorporating the data collection effort into the project scope and time line, and (2) making sure that this need for data collection is communicated to the project champion so that he or she understands and agrees.

The majority of LSS projects are "certification projects," in that the project leader is simultaneously learning new skills, applying them to a problem that no one previously has solved, and managing a project team. Part of the certification process involves the LSS practitioner demonstrating a core set of quality tools on a project, including qualitative tools like process maps and fishbone diagrams, and quantitative tools like control charts and inferential statistics (hypothesis tests). For BBs, the expectations usually include advanced techniques like statistical tolerances, design of experiment (DOE), and components of variation analysis. Any engineer will tell you that problem-solving begins by drawing a picture and that is true for LSS—a process map is a universal tool. But the vast majority of LSS projects can be solved with a process map, a cause-and-effect tool like a fishbone diagram, maybe some simple graphs like Pareto charts or time charts, and a control plan. A complicated project may need a simple individual control chart and a risk assessment tool (failure modes and effects analysis or FMEA) or comparative tool (cause-and-effect matrix or Pugh matrix). Few projects ever require the rigor of a DOE.

Consider an Environmental Health and Safety (EHS) project focused on reducing hazardous waste disposal for a hot-melt adhesive process line, due to leakage. Over 1,000 pounds of hot glue had to be disposed each month (for years!), so the EHS manager found an environmentally friendly recycler to purchase the waste. The process had to be documented and improvements emplaced, to ensure that the waste adhesive was free of extraneous debris. The seasoned GB project leader mapped the process and pulled a team together for an FMEA session. During that session, a process worker pointed out that, years earlier, an engineer switched adhesive filters to save money, going from round to square, and that's about when the leaking began (Yes, I am going there—they tried to fit a square peg in a round hole!). The team immediately ordered some round filters and installed them at the next opportunity, and the leaking was stopped. The GB, with support of a MBB, convinced the plant leadership that the process improvement was in fact a LSS project, despite taking only two meetings and a process map. It was a great achievement that the original deliverable of recycling revenue from the leakage was substituted with a nearly 100% elimination of the leakage itself.

Data, rigor, and management expectations are recurring themes that influence both individual project success and LSS program perceptions, and communication is often a practitioner's most important tool for controlling that influence, as the examples that follow will demonstrate. Please note that actual results, target values, and data are fictional to avoid improprieties, but the projects, communications techniques, and general outcomes are very real.

BBs Are Not Superheroes

The workhorse of the LSS deployment is the BB. These brave souls are often pulled from their current role, provided 4–5 weeks (spread over a few months) of postgraduate level training on process analysis, applied statistics, and quality management-related techniques, and simultaneously challenged to apply that learning to solve a problem that no one prior to them in the history of the business has been able to solve. Fortunately, they are usually given until the end of training plus a month or so to wrap up their work. To further justify the salary premium that may accompany this new job, the BB is expected to demonstrate a wide range of quality management tools on their project, or their work "isn't a Six Sigma project."

For the BB who survives her training project and become certified, the demands often become more impressive. Now, she is expected to "deliver" three to four projects per year, with savings that range from a few hundred thousand to a million dollars per project, while working on tougher problems and leaving the low-hanging fruit for GBs and trainees.

Like most urban myths, these expectations are rooted in some past reality, often tracing back to GE's early successes in Six Sigma. In reality, goals that work for a billion-dollar business may need to be adjusted considerably for a $20 million operation. For the analyst-turned-practitioner, setting and managing realistic expectations is often the most important aspect of communicating with executives.

Case I: I Expect Big Things from Six Sigma—Warehouse Operations

During a Six Sigma rollout at a large retailer, a common tactic for selecting BBs was to identify high-performing, seasoned employees who are likely to move into higher management roles and put them through a rotation as a BB. The business leaders were provided a half-day LSS overview course, so they could better understand the corporate program and help manage their local portfolio of projects.

Such was the case with a general manager who promoted a warehouse supervisor to BB and declared that she expected big things from Six Sigma. The first project was to improve shipping accuracy to 99.95%, a target the general manager learned from a Lean conference that applied to world-class shipping operations. Immediately, the BB had three major objectives: satisfy the technical requirements of the certification, manage the lofty expectations from management, and juggle her previous job requirements until a replacement could be found.

For the technical requirements, the BB followed the well-established path. In Define phase, she created a charter, conducted an initial financial estimate, collected voice of the business and voice of the customer, established process boundaries, defined the defect, identified available data sources, and finalized the project objectives. Measure phase included a series of detailed process maps for the order-pick-pack-ship operation, validation of the measurement system, and a control chart to establish process stability and baseline mean. This phase also included some benchmarking to validate the GM-provided objective, which seemed to apply to brand-new, fully automated warehouses with limited human interaction.

This warehouse was old and very much manual. While the baseline shipping accuracy was already an impressive 99.3%, it became clear very early that a 99.95% goal was unrealistic. The team believed that a goal of 99.6% was achievable given the time and limited resources available, but they were concerned that the GM wouldn't agree to such a small improvement. In reality, moving from 99.3% to 99.6% required a defect reduction of about 50%, and put into tangible terms, that meant eliminating approximately 7,000 shipping errors per year with no capital outlay. From the first two project phases, there were just a few key pieces of information that the executive needed, which are summarized in Figure 8.1a.

Chi-square, statistical process control, multiple process analysis, meetings with process employees, reams of pivots tables, and descriptive and comparative statistics, went into this result, all of which the BB needed to advance the project (and to get the MBB's blessing toward certification). But the executive needed a bottom-line-up-front summary tied directly to business needs and that took a single page.

Figure 8.1b further emphasizes the level of effort associated with the general manager's original, higher percent-accuracy goal and the

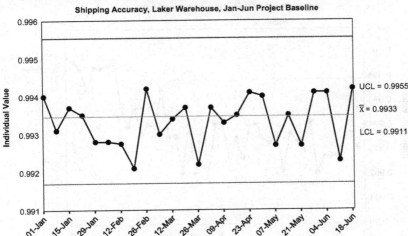

Figure 8.1a   Warehouse operations summary, Measure phase.

Percent Accurate	Defect Percent	Number of Defects
99.3%	0.7%	14,000
99.6%	0.4%	8,000
99.95%	0.05%	1,000

Reducing defects from 14,000 to 1,000 was unrealistic. As a first step, reducing defects to 8,000 was an achievable target.

Figure 8.1b   Illustration of defects as a function of accuracy percentage.

basis for the project leader's successful argument for a more achievable goal. The new project objective was approved.

The Analyze phase required an FMEA, some cause-and-effect tools, and validation by direct observation. Brainstorming with process experts and a small-scale pilot generated significant improvements (the Improve phase). A monitoring plan involving a simple control chart was created (the Control phase). The BB and the MBB met and went through the inferential statistical comparisons, the updated process maps, the revised standard operating procedures, and other technical requirements. For the executive presentation, the focus again was bottom-line-up-front, this time with a visual control chart that clearly illustrated the before and after picture with confidence that the improvements were real (Figure 8.2). The project was

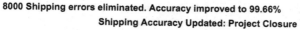

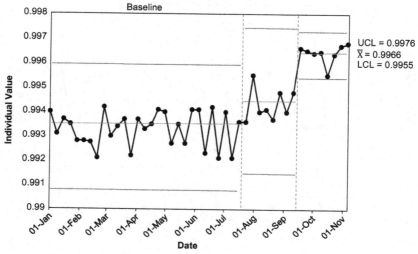

Figure 8.2 Warehouse operations project summary.

closed with a new process baseline greater than 99.6% and more than 8,000 more customers are getting what they ordered.

GBs Have Day Jobs

Another common tactic for LSS deployments is to identify high-performing persons who are embedded in their process roles and train them as GBs so they can apply LSS methods to their everyday jobs. The idea is that as these GBs work their way up the organization, they will take process thinking and quality management skills with them to spread as their influence grows. GBs deliberately keep their "day jobs," and LSS training provides them with an additional skill set to improve the performance of the organization in which they work. It works great, if everyone is on board. It works better if the processes and resources GBs target are fully within their control.

In some organizations, simply using the word *Lean* is a nonstarter. One manufacturing plant decided to implement Lean thinking just a year or so after a major layoff and restructuring—they deliberately avoided the word "Lean" and called their deployment continuous flow manufacturing. Executives in a government agency were not keen on naming their Six Sigma deployment "Six Sigma," so they rebranded it with phrases like "embedded quality." In each case, the deployment was successful, helped by targeted communications.

Such was the challenge for a marketing manager, who had to sell science to "creative thinkers." While direct-mail marketing (the focus of the next story) is being supplanted by digital marketing these days, the early parts of the production process are largely the same for both, with print and ship processes being replaced by code and post online toward the end of the value stream. As such, several of the lessons learned from this small GB project were successfully applied to a very large marketing operation years later.

Case II: We Are Too Specialized for Six Sigma— Marketing Materials Production

A larger retailer prepared print catalogs and tabloids several times per year, to be mailed to thousands of customers across the country. Every function in the marketing organization was at full capacity, with backlogs of work and a relentless sense of urgency. A very bright manager was chosen for GB training and challenged with reducing the cycle time for product development. It took an average of seven weeks to complete a marketing campaign to create a tabloid, and eight weeks to create a catalog, and everyone in the organization sensed that there was a plenty of waste to be eliminated. The GB had two particular challenges to realizing these gains: everyone thought that the problem was in someone else's area, and the creative services employees in particular believed that Six Sigma simply did not apply to them, since creativity cannot be put on a time line. The common perception was that creative services was the driving force behind the marketing effort, and without their cooperation, the project seemed doomed.

The GB worked through the Define and Measure phases with much the same analytical tools described earlier. For this project, however, the focus was time (the project objective was a 10% reduction in production time), which necessarily meant value-added analysis was important. A great deal of detailed process mapping (coupled with a sense of humor and some appeasement to groups who took offense to their job being labeled "required waste") and time studies were done. The data collection was particularly difficult because the timescale for a marketing campaign was many weeks, and little reliable contextual data were collected and retained on any specific aspect of the campaign (e.g., reasons for delays and rework were not explicitly documented and tracked once a product was finished). Fortunately, all work products were date stamped, so a clear picture of the current baseline became evident after some manual data analysis effort, and it was revealed that more than 50% of total production time was spent in reviews and changes, a classic rework loop. While it was an eye-opener for the executive, it meant that significant savings could be gained by focusing on these areas and leaving the creative types alone. In addition, the measurement system analysis and data collection plans were the first attempts by this organization to formally track defects. They identified nine types of defects and determined that the average product had more than a dozen defects across all types (e.g., wrong price and wrong product description) Figure 8.3 represents the results of the analysis that the GB communicated to the marketing executive championing the project.

The most common defect types were in the activities of layout and product description, and the longest processes in a typical campaign

Marketing Production Process Baseline Findings

- Process baseline, avg days to complete:
  - 55 days, tabloid
  - 65 days, catalog

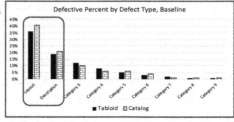

- Process baseline, errors:
  - 40% in style and layout
  - 20% in descriptions

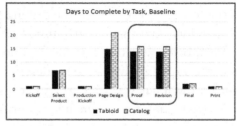

Project Goal: 10% reduction in total production time per product

Figure 8.3 Marketing production summary, Measure phase.

were proofing and revision. Specifically, proofing and revision activity times could be reduced by eliminating or reducing layout and product-description errors. What is most important here is that the executives were only presented with the results as they pertained to the business. The details of the various analyses performed were not included in the presentation.

The Analyze phase for this project involved tedious evaluation of many print samples from previous production runs. Each defect was manually traced back to its source. An FMEA, coupled with multiple reviews and revisions to detailed process maps, eventually revealed one root cause of the problem. It was revealed that creative services pulled merchant information from a particular database, but merchants updated a different database, so the one that marketing used was often not current. Many of the changes to the product directly resulted from this disconnect. For a GB to affect a database owned by another business function was beyond the scope of the project, but the information was passed along for a long-term, permanent fix. In the meantime, a simple fix was implemented to catch product discrepancies up front, and several other process improvements were made including documenting standard practices for the first time. Interestingly, the creative group that felt Six Sigma didn't apply to them were the beneficiaries of the standard-practice documentation, without once being told to do their job differently. Pilot results from the Improve phase yielded a 25% reduction in defects and a 33% reduction in proofing and change time, which resulted in an overall production time savings of 10%, the original project goal. Figure 8.4 represents the culmination of several months of investigation, analysis, and control planning, presented to the business executive.

The GB certainly could have thrown in very complex flow diagrams, and dense tables of summaries and comparisons. However, a couple of bar charts and some bullet points successfully communicated the results.

Do's and Don'ts of Presenting LSS Work to Leadership

LSS practitioners normally find themselves briefing executives under two broad conditions: project reviews and program reviews.

Marketing Production Project Closing Summary

- Improvements, errors:
  - 25% error reduction

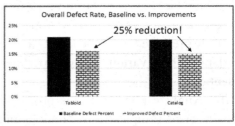

- Improvements, avg days:
  - Tabloid, 5 days saved
  - Catalog, 6 days saved

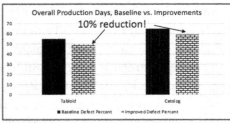

Overall 10% reduction in production time (project objective reached!)
Improvements: 25% error reduction, 33% time reduction in Proof & Revision

Figure 8.4 Marketing production project summary.

Each briefing has a specific purpose and requires unique preparation, but success for either meeting involves similar preparative guidelines.

The LSS practitioner should keep in mind that the business executive wants to see the business impact; the detailed analyses are normally intended for a MBB's technical approval. A single-page executive summary should contain exactly enough information for an executive to make a decision to move forward, accept project deliverable changes, or stop the project. Even if a review is scheduled for 30 minutes, it is best to assume you have 5 minutes to make your case, and the remainder of the time is just an opportunity to showcase good work.

Program reviews involve many projects being summarized, and if they are part of a C-level review, the entire program update may be distilled to a single executive summary page. Here, details are distractions. The project leader needs to be clear and concise, and state exactly what the benefit to the business is, when it will be delivered (if the project is not closed), and any cautions or barriers that need to be elevated to that level of leadership. "No surprises" is a critical mantra for program reviews. If there is bad news, it should have been delivered and a resolution sorted out before the program-level review. Even good news should not be a surprise. Don't let an executive ask, "Why didn't I know this sooner?"

Conclusion

For LSS practitioners, communication is a skill as important as the technical skills in applying process-improvement tools. The success of a project depends on clear communication of the objectives up front, continued communication with the project team and management during each phase of the project, and final communication of the results to the champion and other stakeholders. For any presentation, an analyst must understand the needs of the audience and avoid the temptation to provide the details of every analysis performed. Generally speaking, follow the adage that less is more.

9

INTERNAL AUDITING

Seeking Action from Top Management to Mitigate Risk

JASON R. THOGMARTIN

Santander US

Contents

Introduction

Internal auditing is a field where data analytic techniques have become a key tool to provide assurance on risk, control, and governance processes. For those who may not work with the audit function frequently, it is useful to know that internal audit has been defined by the Institute of Internal Auditors as "an independent, objective assurance and consulting activity designed to add value and improve an organization's operations. It helps an organization accomplish its objectives by bringing a systematic, disciplined approach to evaluate and improve the effectiveness of risk management, control, and governance processes[1]." The collection, aggregation, and analysis of data, and the communication of outcomes form the essence of the evaluation and improvement of effectiveness that this mandate speaks of.

Successful data analysis techniques combined with the effective communication of analytical outcomes is essential to the success of the auditing function and more broadly the organization that is intended to benefit from audit's work. While this chapter discusses

---

[1] https://na.theiia.org/standards-guidance/mandatory-guidance/Pages/Definition-of-Internal-Auditing.aspx.

techniques for the effective communication of analytical outcomes, the concepts and techniques for communication covered here are relevant to any risk and control activity. In the following sections, I first briefly outline how data analytics are used in an audit context. I then present the effective communication strategies and techniques that I used in three specific cases—on risk assessment, cash control, and maturity assessment.

Use of Analytics in Auditing

Data analytics in auditing differs in purpose and expected outcomes, from the analysis techniques deployed in other business areas. The methods used in many cases are not unique to audit, but the direction of use and application is. Audit analytics in this regard is indirect in that it enables the analyst to communicate on risk and control to senior leaders of a company a compelling cause to action to strengthen control, mitigate risk, drive enhanced organization efficiency, and most importantly protect the organization.

Traditional manual audit analysis was limited by individual capacity and used smaller samples and a prioritized list of tests to perform. Today, access to data and computing power enables detailed testing through the evaluation of larger sets of data, the ability to conduct multiple lines of testing, and reaching areas that may not be easily accessed through traditional testing. In the most ideal examples, advanced techniques and computer-assisted analysis provide greater automation efficiency and reduced manual efforts. However, data analytics cannot work in all places. In the worst case, flawed data and flawed analysis call for even more manual intervention to make sense of what was done if anything are to be salvaged.

The benefits realized through analytics are not limited to audit but ripple through the entire business. However, as the cases that follow will show, the impact of the analytics performed is very much tied to the effectiveness of the methods by which it is communicated within the audit department and to key stakeholders at the management or board level. Ineffective communication by audit could allow risks to go unmitigated and actions to not be taken that would improve and enhance processes and efficiencies in the company. That said, regardless of whether analytics or simply traditional procedures are used, communicating simply and clearly to the executive management of a firm is essential. Audit recommendations provide awareness, and support decision-making and actions. Both simple and complex topics communicated with appropriate clarity will ensure that the messages are delivered as intended.

The three cases here show that successful presentations do not require stakeholders to understand the minutiae of the analytic procedures. What is important, however, is to link the results of the analytical processes employed, using simple and straightforward terms, to the business outcomes most relevant to the company. This is easier said than done. Developing and delivering a successful analytic communications strategy requires not only a well-designed and effectively executed analysis but also an understanding of the strategy, priorities, organizational context, motivations, and anticipated reactions, to achieve the desired outcomes. The tenets of tailoring the communication to the audience hold true, and in the case of senior management audiences, care must be taken to match the content to their level of interest and technical knowledge to achieve an ideal outcome. In the cases discussed here, the communication is geared towards senior level stakeholders that have little interest in the specifics of the analysis other than to gain assurance that it was thoughtfully done.

Case 1: Risk Assessment

Description

Risk assessment is the foundational activity for an internal audit function where data are used to gain a comprehensive understanding of a firm's risk profile for the development of appropriate audit coverage. Risk assessment is a key tool within the audit function that is used to set the plan of audit activities for the subsequent year. As such the outcomes are shared with company staff, management, and the board for awareness and alignment around the future activities to be performed. It therefore needs to be easily understood by the members of management and the board. To arrive at a view on risk, a large amount of data must be gathered from across the firm. In the example discussed here, both quantitative and qualitative data are collected from top down (assessments which focus on broad and thematic areas of risk) and bottom up (assessments which focus on granular risk at the product, sub-business area, or functional level). The data are interrogated to arrive at a view of risk that then informs the audit plan for the subsequent period. The organization discussed in this example is large, diverse, and complex across multiple product and geographic lines. As a result, the data produced are quite extensive and require significant analysis through a structured methodology to distill into an actionable format

Analytical Techniques

The data gathered in this project include financial information, prior audit results, outcomes of internal assessments, strategic planning documentation, lists of key projects and initiatives, external and competitor information, etc. The risk assessment methodology is then applied to the data to derive a view of risk by the component entities of the company, for example, product operations, accounting, and regulatory compliance. These data are used to assess an inherent risk as well as a rating on the controls that mitigate those inherent risks. The combined calculation produces a residual risk for each entity which will then inform the frequency and nature of the coverage by the audit function. This risk rating is not only derived mathematically but also adjusted based on qualitative factors that may not be captured in the model. In short, this process can be very complex, producing a significant amount of data on the risk environment for the company. While understanding the nuances of the methodology and the rating model is key for the audit department, it is not so for the management. Thus, it is critical for audit to be able to convey implied assurance that the process is robust without focusing on the details so the discussion is on the outcomes and the proposed actions.

Communication of Results

There are three target audiences for the risk assessment results: (1) within the audit function, (2) company management, and (3) the board of directors. Each of these three have a distinct set of expectations that should be considered in the communication process.

Communication within the audit department revolves around the calibration and challenge of the risk ratings process. Prior to concluding on a proposed audit plan and presenting to management, internal dialogue should occur to solidify the strength of the overall analysis. Given the large amount of data, further data analytics techniques are required to reasonably aggregate and analyze information for understanding and action. Figure 9.1a illustrates the kind of information that the internal audit team would find useful.

For company management, data visualization is a useful tool to make the numbers more meaningful. A chart such as the one in Figure 9.1b shows how the tabular data used for detailed analysis can be pivoted into a series of bar charts and graphs to enhance comparability across variables and better illustrate outliers.

The detailed, aggregated data and the visualizations work well for the presentation of the analysis to the internal audience. It supports the internal team in assessing the appropriateness of the overall outcomes. The charts in this case are descriptive, and it is left to the management to make a decision on the strategic direction as a result. The contextual knowledge of the business function that is being analyzed, possessed by the internal team, plays a key role in their ability to make sense of these charts. However, these charts may not be suitable for external stakeholders.

External audiences do not always have a desire or need to sift through large amounts of tabular data and think about whether it makes sense and the resulting outcomes are reasonable. This consumes time and effort that is best spent elsewhere by these stakeholders. Therefore, it is critical for the auditor to present an actionable summary of outcomes. Figure 9.1c provides such a summary in a two-dimensional (inherent risk vs. current control environment) matrix for the units within the Global Manufacturing Business.

The chart above is powerful as it quickly allows the audience to see visually where risk is concentrated based on audit's assessment. For example, product line A is assessed to have high risk and a high level of existing control, while product line D is low risk and has a low level of existing control. This match of risk and control indicates a satisfactory state of affairs. In other words, anything along the diagonal of the matrix, from the lower left to the upper right indicates that the risk management in place is satisfactory. Anything that goes away from the diagonal requires further attention. Specifically, units

(a)

Units within the Global Manufacturing Business

Auditable Unit	Interest Risk	Adjustments	Adjusted Inherent Risk	Control Environment	Adjustments	Adjusted Control Environment	Residual Risk	Adjusted Residual Risk	Overall Risk Level
Sales	3.08		3.08	1.03	-0.5	0.53	2.05	2.55	Moderate
Operations	2.05	-0.5	1.55	1.03		1.03	1.03	0.53	Low
Marketing	2.05		2.05	1.03	0.5	1.53	1.03	0.53	Low
Legal	2.05		2.05	0.00		0.00	2.05	2.05	Moderate
Human Resources	4.11	0.5	4.61	1.03		1.03	3.08	3.58	High
Purchasing	3.08		3.08	2.05	-1.00	1.05	1.03	2.03	Moderate
Finance	3.08		3.08	1.03	0.5	1.53	2.05	1.55	Low
Accounting	4.11	1.00	5.11	2.05	-0.5	1.55	2.05	3.55	High
Communications	4.11		4.11	1.03		1.03	3.08	3.08	High
Technology	3.08		3.08	0.00		0.00	3.08	3.08	High
Product Line A	2.05		2.05	2.05		2.05	0.00	0.00	Low
Product Line B	2.05	1.00	3.05	0.00		0.00	2.05	3.05	High
Product Line C	2.05		2.05	2.05		2.05	0.00	0.00	Low
Product Line D	4.11		4.11	1.03		1.03	3.08	3.08	High
Product Line E	3.08		3.08	1.03		1.03	2.05	2.05	Moderate
North America	2.05		2.05	0.00	1.00	1.00	2.05	1.05	Low
South America	2.05		2.05	1.03	1.00	2.03	1.03	0.03	Low
Europe	3.08		3.08	1.03	1.00	2.03	2.05	1.05	Low
Asia	3.08		3.08	1.03		1.03	2.05	2.05	Moderate

(b)

Businesses and functions within the Overall Company

Categories: Global Manufacturing, Global Services, Global Distribution, Research Center, Shared Service Center, Information Technology, Finance, Human Resources, Administration

Legend: Low   Moderate   Elevated   High   —— Planned Audits

(c)

Units within the Global Manufacturing Business

Inherent Risk (vertical axis) × Control Environment (horizontal axis); columns: Low, Moderate, High

	Low	Moderate	High	
High	Europe	Asia / Product Line B	Product Line A / Technology / Geography XX	High
Moderate	Accounting / Finance	Purchasing / Sales	Product Line E / North America / Human Resources	Moderate
Low	Product Line C / Product Line D / Communications / Operations	Geography XX / Function XX / Marketing	Legal / South America	Low

Key

In Plan
Strategic Priority
Less than Satisfactory
areas shown in Bold

This chart visualizes the risk profile for the global manufacturing business in Fig. 1a. Note the risk of each function, geography, or product have different risk attributes in this figure for illustrative purposes. The nine block enables the reader to visual risk by additional modification to the font or text color conveys additional information without having to reference a separate document. For example, the reader knows Technology has a high residual risk, was previously evaluated as less than satisfactory, and is in the current audit plan.

Figure 9.1   Risk assessment.

at the bottom right of the matrix are ones that are over controlled, and resources can be freed up from there. Likewise, units on the top left are under-controlled and need greater risk mitigation strategies in the future.

Case 2: Cash Controls

Description

In this case, I present a more traditional operational auditing area that involves controls around a company's cash management processes. Many analytic techniques are used in the day-to-day execution of audits such as the analysis of accounts payable, travel expenses, inventory controls, and others. As in the previous case, the broad objective of the audit is to evaluate the degree of risk present and discover weaknesses in the control mechanisms. Specifically, the purpose of the cash control audit is to analyze and determine whether improper movements of cash have occurred, remediate issues, and strengthen controls to prevent reoccurrence. Traditional audit techniques in cash controls involved manually reviewing only a small sample of the data and running the risk of potentially missing transactions that do indeed pose a risk. Using analytical techniques, one can now evaluate the entire population of transactions by connecting key sources of data in treasury systems, bank accounts, and ledgers and analyzing these data to find certain patterns and trends that would be potential red flags for improper activity.

Analytical Techniques

The data produced in this project included all the details surrounding movements of cash within the firm. Transaction details were extracted from all bank accounts systems and related source documentation noting the amount of money moved, parties to the transaction, request and processing dates and times, physical locations of requestor and recipients, requestor and approver of the transaction, etc.

With the inventory of data elements secured, specific scenarios for testing and monitoring of cash movements were developed where outliers could represent improper activities. Examples of scenarios included confirming proper approval levels and persons, checking for duplicate payments, and checking unusual activities such as timing of payments out of order based on historical trends, changes in bank accounts where cash was transmitted, payments always staying below certain policy approval thresholds, and unusual activities by geography or time zone. With the help of the data, these scenarios are checked for possible red flags. The issues thus flagged are candidates for manual review to confirm or clear. One potential problem with the analytical techniques could be the creation of too many false positives (red flags)—for scenarios that are in fact in control. Excessive erroneous red flags drive significant manual work and potential distraction from actual issues that may be found otherwise. This requires that the analytical methods used are reliably accurate. The internal management team wants to ensure that the methods used as well as the outcomes are reliable (scenarios with red flags). However, an external audience is only interested in the final outcomes.

Communication of Results

Similar to the previous case on risk assessment, the presentations are structured based on the needs of the audience. For the internal audience, the tactical details are presented and assumes a certain level of understanding and familiarity with the analytical techniques. What worked well for the internal audience in this case was more granular dashboards showing the progression of the analytical approaches over time. As the analytical exercise itself was a multistep project, communications covering the entire process was key. This involved providing a detailed grounding on the analytical methods, the sources of the data, and ability to secure the data, along with overall data quality and the methodology to be employed in designing the various scenarios. Figure 9.2a illustrates an interim update presented during the execution of the analytical exercise.

The focus here is on project management for the analytical exercise. Due to the need to detail the multistep analytical process, a large number of documents were created. The challenge to the audit staff is in determining what documentation needed to be presented, so that the presentation can be better focused. Such presentations covered initial methodology and project approvals, interim dashboards showing progress for the analytics program, and final outputs showing effectiveness of each scenario tested and outcomes that are being rolled up for presentation to management.

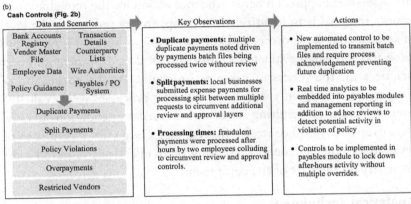

(a)

Cash Controls (Fig. 2a)

Scenario	Status	Population Reviewed	Red Flags	Review Progress	Confirmed Errors	Adjustments	Next steps
Duplicate Payments	→	87,548	1,575	75%	7	Logic error produced red flags for recurring pmts, analysis re-tuned to remove.	Complete red flag reviews, confirm exceptions
Split Payments	→	247	12	65%	1	None required	Complete testing
Bank Account Set-up	√	595	2	100%	None	Script missed bank accounts for corporate segment	Add new accounts to population, complete testing
Counterparty Approval	→	1,124	45	95%	3	None required	
Vendor Approval	→	788	108	50%	5	Some vendors were duplicated in the population	Remove duplicate vendors
Timing	→	47	21	15%	2	Payments in Asia analyzed based on EST producing enormous red flags	Rerun Asia payments using correct time zone

(b)

Cash Controls (Fig. 2b)

Data and Scenarios	Key Observations	Actions
Bank Accounts Registry / Transaction Details Vendor Master File / Counterparty Lists Employee Data / Wire Authorities Policy Guidance / Payables / PO System Duplicate Payments Split Payments Policy Violations Overpayments Restricted Vendors	• **Duplicate payments:** multiple duplicate payments noted driven by payments batch files being processed twice without review • **Split payments:** local businesses submitted expense payments for processing split between multiple requests to circumvent additional review and approval layers • **Processing times:** fraudulent payments were processed after hours by two employees colluding to circumvent review and approval controls.	• New automated control to be implemented to transmit batch files and require process acknowledgement preventing future duplication • Real time analytics to be embedded into payables modules and management reporting in addition to ad hoc reviews to detect potential activity in violation of policy • Controls to be implemented in payables module to lock down after-hours activity without multiple overrides.

Figure 9.2 Cash controls.

For the external audience, a different approach was taken. The focus was not on the data or the tactics behind the analytical exercise but rather on the story that the data tell. Senior stakeholders are not usually interested in data for its own sake but what it can tell them about the facts and circumstances concerning the business—in this case, illustrating where cash has been improperly moved or handled in the company…a key concern! Figure 9.2b shows a summary of the specific evaluation of the outcomes and call to action.

While the above presentation plans for a very specific profile of senior leadership, it oftentimes doesn't happen as planned. The overriding goal in the examples is to remove as much complexity and distraction and let the data tell a story. Ideally, this is the story that you want conveyed to the audience. However, oftentimes there is someone who wants to dig deeper, which means being prepared with details in the form of supplements or appendices is always wise.

Case 3: Multi-Business Line Controls Maturity Assessment

Description

The business case for maturity assessment arises when traditional audit approaches stop being effective. For instance, if it was already established that controls were weak for key processes across businesses in a certain region, then continuing to apply a traditional audit methodology would conclude what was already known. The maturity assessment approach helps influence action for the remediation of issues. The goal is to identify specific gaps in key processes, the level of associated risk exposure, and key expectations on what is needed to close the gap when evaluated against the local environment and best practices for the processes examined globally. At the conclusion of the project, over twenty significant business processes were evaluated over six countries, resulting in reviews and recommendations to close gaps over nearly 500 controls. This created a large set of results to be presented to management. Presenting them effectively required some creativity on the part of the audit team.

Analytical Techniques

The data produced in the project consisted of information on key processes, standardized sets of questionnaires applied across geographies including responses, control documentation across every control reviewed, best practice expectations for standard controls expected, documentation of gaps found aligned to controls, and narrative information on risk exposures with recommended prioritization of gaps to be closed. In addition, to further drive understanding on the overall environment and assist both local and senior stakeholders in understanding outcomes, the data produced during the audit needed to be further synthesized into aggregate views. Thus, new insight was created summarizing the overall maturity of processes by country and region.

Communication of Results

As mentioned above, a robust synthesis of the data is critical for effective understanding of the audit project and its outcomes. At the end, the data are not useful to senior stakeholders if they do not tell a story. The outputs from this maturity assessment present a call to action at both the local and the corporate levels. If effectively presented, the data will clearly show where organizational priorities need to be focused to enhance controls, mitigate risk, and better protect the company.

To make an effective presentation to top management, the audit team had to decide how to distill and present the large volume of results in a way that does not overwhelm the audience, while retaining enough detail to make it relevant for the development of specific action plans. Figure 9.3a illustrates one of the slides used to present to the local management team. These are the people who are accountable for the processes will be responsible for putting in place any fixes or remedial actions required.

Thus, it was very important for them to see the discrete issues and specific areas of action required to better control the key processes examined in their business lines. A key aspect of this output was the context of making the business better and protecting the company. By

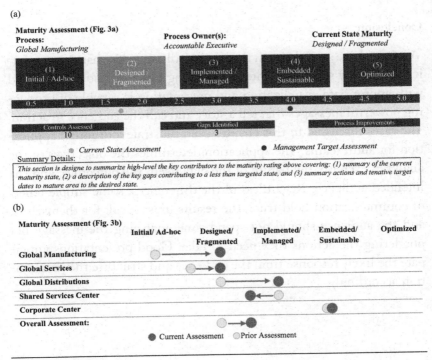

Figure 9.3 Maturity assessment.

showing what needs to occur to reach a mature process, both audit and the local members of management are aligned on the same side which produces a collaborative dialogue compared to a traditional report on audit outcomes that is seen in the second project described above.

The senior level presentation is cast differently. Instead of providing data at the level of granularity found with the local users, it has been summarized in a maturity matrix (Figure 9.3b) visualization.

Gone are the notations of individual gaps and weaknesses but present here is an overall illustration of the maturity by country, business, and process. When compared with charts for other areas, a baseline can be established and senior stakeholders are able to understand and trend risk across the company. Also of note is while this format is targeted to senior stakeholders, it was also shared with local management as well. While more granular information is useful in the tactical local actions that will need to occur, it is also important locally to understand how the information will be presented throughout the company and see the broader context the outcomes are being placed into.

Conclusion

What we have seen through the projects discussed is that data analytics plays a key role in the internal auditing function and when paired with appropriate visualization and presentation is capable of producing powerful messages that tell a story and call the audience to action. The cases discussed in this chapter each illustrate a different application for audit teams. The application determines the objectives of the analytics exercise, the nature of the data itself, the outcomes, and the intended audience. Regardless of the differences, some simple tenets of communication hold true. The results must speak for themselves, and the audience should not spend time analyzing the message but pondering the actions they need to take. Good presentations anticipate the likely response from the audience and structure the presentation accordingly.

10

CONSUMER LENDING

Winning Presentations to Investors

J. P. JAMES
Lending Science DM

Contents

The Backdrop

Early in 2017, I was asked by an acquaintance whose company was recently acquired for my thoughts on what types of businesses might be good early-stage investments with moderate risk exposure, as I had almost two decades of experience building companies and managing investments. During our discussion of different business models, I mentioned an idea for an online consumer lender. Having helped build a marketing agency in the space and then acquiring a credit scoring company, I knew that even in the big data era, many lenders were not truly leveraging the tools of data science: analytics, machine learning (ML), software development, and scalable IT infrastructure. We continued talking through the general risks of new ventures along with how I might recruit a team that possessed core competencies in underwriting, operations, and capital raising to build a successful lender. With his support and a seed round of capital, I assembled the team, and in less than six months, we were lending. With this rapid execution, we attracted both private and institutional investors, springboarding us to quickly capture significant market share in the online consumer lending marketplace. The obstacles we faced along the way underscore the need for focused, outcome-oriented business analytics, and data management. This is a discussion of how data, software development, and analytics have been at the core of our rapid expansion and management of risk.

Building the Systems

An intricate web of IT architecture and application programming interfaces were necessary to manage both our lending operations and the data flow for analysis. Though I have experience building businesses across a variety of industries, starting a lender was unique in the complexity of this architecture. The systems requiring implementation included the following: the lead channels, digital marketing and underwriting processes, the loan origination and management platform, multiple payment processors, collections, and remarketing systems. These systems had to talk to each other while pulling the data into a single repository for analysis. We used process maps similar to the one below to visualize their interconnectivity and identify all data capture points throughout the entire loan origination and customer management cycle. "Future state" maps featuring automated processes could then be built on this foundation along with the data capture implications for ML.

ONLINE LENDER PROCESS FLOW

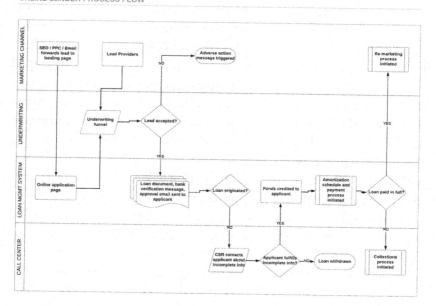

Infrastructure and reporting across functional areas were developed in an "agile" framework, popular in software development for rapid deployment of a product in an iterative fashion (compared to traditional project management). We used Asana (a project management software) to centralize team information, communication, and responsibilities, and created project dashboards identifying primary outcomes for each area (IT, finance, business intelligence, operations, etc.) that were then broken into clearly defined subtasks with completion dates. Management and investors could refer to the dashboards for progress updates as we got underway, while the operations team had a prioritized project listing that new team members could also use for orienting themselves with our business objectives. From a project standpoint, this information hierarchy drastically reduced organizational complexity.

Due to security requirements and speed of execution at scale, we chose a large, well-supported architecture and complimentary programming languages allowing us the flexibility to pull data into other systems for analysis. We worked with providers to feed the latest regulatory language and protocols into our loan documents, and specialized law firms to spot-audit our processes. Our reporting tools were integrated with green light checkoffs that showed if the audits were successful or not. Making these decisions early on was important as the architecture was designed to support our analytics tools while not handcuffing the organization to a technology stack that could very well become outdated within a year. We were also positioned to efficiently implement ML algorithms across the business (marketing, underwriting, and collections) in the future once our data were at scale.

Audience Drives the Reporting Needs

The point of focusing on data capture was to serve the reporting needs of our investors, management, and operators. Fundamentally, these three audiences (should) drive the analytics efforts for every organization. Analysts need to understand this—too many businesses are sitting on mountains of data they don't know what to do with or are producing reports with no clear impact on the organization at large; the analysts know the "what" but not the "why" for their output.

Investors are looking for growth opportunities, measured risk, and outsized returns. Any report oriented towards them needs to address these topics while staying attuned to the varying levels of investor sophistication: have they invested before? What is their risk tolerance? Do they understand the industry or business model? Do they have issues with the industry? This informs the investor's psyche. They don't want a data dump of tables with no order or emphasis, and they certainly shouldn't have to dig through operational minutiae to know if the business is performing and they will get a return. The narrative and visuals should illuminate performance such that they feel confident in their decision and want to invest more.

Management needs real-time performance metrics across the organization's functional areas for *efficiently* making decisions and determining overall strategy. Key performance indicators must be contextualized against benchmarks or within performance bands that imply clear courses of action. Similarly, the operations team needs actionable metrics, so they can prioritize their output and adjust performance as needed. The reporting needs to be focused and free of unnecessary detail while communicating each metric's importance. Again, we built the lender (and its data infrastructure) with this output in mind.

Even with the experienced management team, it required tremendous research to actually understand the financials of consumer lending and how our specific business model would work with all of the associated costs. Marketing strategy in lending comes down to cost per acquisition, time per acquisition, and channel scalability, which in turn relates to customer risk and expected lifetime value—the customer's projected profitability across various default, collections, and renewals scenarios. Our projections, however, would have to be built on generic models informed by the team's industry experience and vendor benchmarks until we had enough data to build our own. We used cost and conversion benchmarks from the team's digital marketing background as the foundation of our growth strategy, and reached out to multiple lead providers and underwriting partners to help define the dimensions of the marketplace.

Analytics, Visualization, and Storytelling

With the systems in place, we needed to start generating revenue while gathering the good and bad performance data needed for credit modeling and ML. This depended on month-over-month growth in originations along with managing the first payment default (FPD) rate—loans that were immediately delinquent with no recovery of principal or interest. We displayed our incredible growth over the first 6 months of lending to investors in a chart similar to the one below:

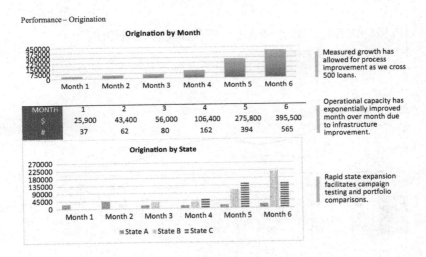

Performance – Origination

Origination by Month

MONTH	1	2	3	4	5	6
$	25,900	43,400	56,000	106,400	275,800	395,500
#	37	62	80	162	394	565

Measured growth has allowed for process improvement as we cross 500 loans.

Operational capacity has exponentially improved month over month due to infrastructure improvement.

Origination by State

Rapid state expansion facilitates campaign testing and portfolio comparisons.

■ State A ▦ State B ≡ State C

Though investors would immediately see the green originations chart at the top, they would focus on the central table emphasizing dollars lent; this was really the most important information for investors, as more money out meant more payments in. Monthly originations were broken down by state as non-stacked bars to maintain visual continuity with the top while facilitating quick comparison of growth by state. The sidebar provided context without being overly detailed—when presenting, I could expound or clarify the points as needed. Besides visual flow, I stressed the balance of graphs, tables, and text on a given slide: it is easy to wow investors with pretty bar charts, but they need some numbers backing them up for emphasis and some bullet point takeaways.

From an operational standpoint, the FPD rate was actually the most important metric. Our entire portfolio was at risk if it rose to unsustainable levels, but even incremental increases into certain lower default bands risked adverse action from our vendors. Investors understandably might read the situation as "FPDs must be as close to zero as possible," resulting in a completely risk-averse approach severely limiting growth, but in fact, a certain percentage of defaults was needed to build a profitable, scalable lender in the first place; the counter-intuitiveness of low volume/low risk not equaling greater profits had to be carefully explained to investors, so they wouldn't get cold feet when looking at early performance data.

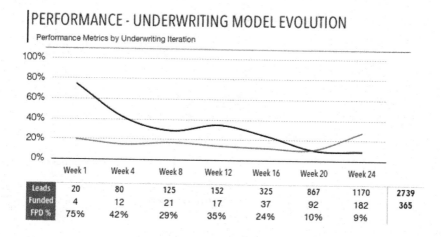

PERFORMANCE - UNDERWRITING MODEL EVOLUTION
Performance Metrics by Underwriting Iteration

	Week 1	Week 4	Week 8	Week 12	Week 16	Week 20	Week 24	
Leads	20	80	125	152	325	867	1170	2739
Funded	4	12	21	17	37	92	182	365
FPD %	75%	42%	29%	35%	24%	10%	9%	

The above chart is similar to what we presented to investors, comparing the FPD rate to the funding rate for a typical early-stage lender[1]. With our unique, rigorous marketing and underwriting processes, we were looking at lower overall funding rates that would incrementally improve over time against initially high default rates that would stabilize within a tolerance band below 40% due to operational optimization—again, this was necessary for model building. Process and system optimization was not going to, nor was it intended to, totally wipe out defaults.

Our internal reporting dashboard examined the FPD rate from a number of angles, including lead provider, state, and loan size distribution breakdowns:

---

[1] For reasons of confidentiality, I have not included our actual performance data for the following charts.

| Loan Count | Origination Month | | | | | |
FPD Status	Group 1	Group 2	Group 3	Group 4	Group 5	Grand Total
1	21	19	25	37	45	147
0	15	47	63	188	436	749
Grand Total	**36**	**66**	**88**	**225**	**481**	**896**

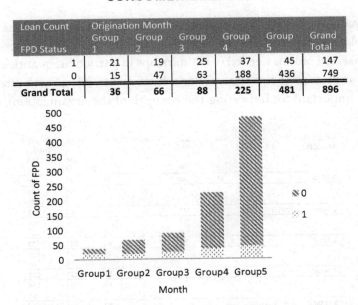

| Loan Count | Origination Month | | | | | |
FPD Status	Group 1	Group 2	Group 3	Group 4	Group 5	Grand Total
1	58.3%	28.8%	28.4%	16.4%	9.4%	16.4%
0	41.7%	71.2%	71.6%	83.6%	90.6%	83.6%
Grand Total	**100%**	**100%**	**100%**	**100%**	**100%**	**100%**

Stacked bar charts are less elegant than the performance metrics curves of the previous slide—investors could instantly follow each curve's movement in relation to one another—but from an operations standpoint, it was more meaningful to see FPDs directly as a percentage of active loans for each month. Whereas investors might be less concerned with the actual quantities of originations and FPDs, internally these were always presented alongside the percentages for a complete picture of the scale of change. For investors, quantities were emphasized only as needed (with start-ups in particular, quantities are great as performance milestones or achievements, but growth rates are more important for indicating the strength of the organization).

Originations Provider	FPD	Month Group 1	Group 2	Group 3	Group 4	Group 5	Grand Total
1	0	5	14	18	33	80	150
	1	12	9	3	8	4	36
1 Total		17	23	21	41	84	186
2	0	9	15	20	48	81	173
	1	9	6	7	10	8	40
2 Total		18	21	27	58	89	213
3	0	1	18	15	34	97	165
	1		4	12	6	5	27
3 Total		1	22	27	40	102	192
4	0			8	28	87	123
	1			1	9	18	28
4 Total				9	37	105	151
5	0			2	45	91	138
	1			2	4	10	16
5 Total				4	49	101	154
Grand Total		36	66	88	225	481	896

Originations Provider	FPD	Month Group1	Group2	Group3	Group4	Group5	Grand Total
1	0	13.89%	21.21%	20.45%	14.67%	16.63%	16.74%
	1	33.33%	13.64%	3.41%	3.56%	0.83%	4.02%
1 Total		47.22%	34.85%	23.86%	18.22%	17.46%	20.76%
2	0	25.00%	22.73%	22.73%	21.33%	16.84%	19.31%
	1	25.00%	9.09%	7.95%	4.44%	1.66%	4.46%
2 Total		50.00%	31.82%	30.68%	25.78%	18.50%	23.77%
3	0	2.78%	27.27%	17.05%	15.11%	20.17%	18.42%
	1	0.00%	6.06%	13.64%	2.67%	1.04%	3.01%
3 Total		2.78%	33.33%	30.68%	17.78%	21.21%	21.43%
4	0	0.00%	0.00%	9.09%	12.44%	18.09%	13.73%
	1	0.00%	0.00%	1.14%	4.00%	3.74%	3.13%
4 Total		0.00%	0.00%	10.23%	16.44%	21.83%	16.85%
5	0	0.00%	0.00%	2.27%	20.00%	18.92%	15.40%
	1	0.00%	0.00%	2.27%	1.78%	2.08%	1.79%
5 Total		0.00%	0.00%	4.55%	21.78%	21.00%	17.19%
Grand Total		100.00%	100.00%	100.00%	100.00%	100.00%	100.00%

FPDs broken down by marketing channel were displayed as heat maps, similar to above, where a cold-to-hot color spectrum represents categorical data. The left chart emphasized monthly totals relative to quarterly totals on a softer color spectrum to highlight months of note. The percentage breakdown used the same legend colors as the investor chart so the team could immediately single out any concerning months or lead providers. As our domestic operations scaled, we could then transpose them to investors as geographic heat maps for dynamic state performance comparisons.

Bin	400-499	500-599	600-699	700-799	800-899	900-999	1000-1099	1100-1199	1200-1299	1300-1399	1400-1499	1500-1599	1600-1699	1700-1800	Grand Total
1	8	10	7	11	17	7	15	7	10	14	5	12	9	15	147
0	58	48	74	55	44	50	51	41	50	51	62	55	55	55	749
Grand Total	66	58	81	66	61	57	66	48	60	65	67	67	64	70	896

Bin	400-499	500-599	600-699	700-799	800-899	900-999	1000-1099	1100-1199	1200-1299	1300-1399	1400-1499	1500-1599	1600-1699	1700-1800
1	12.1%	17.2%	8.6%	16.7%	27.9%	12.3%	22.7%	14.6%	16.7%	21.5%	7.5%	17.9%	14.1%	21.4%
0	87.8%	82.8%	91.3%	83.3%	72.1%	87.7%	77.3%	85.4%	83.3%	78.5%	92.5%	82.1%	85.9%	78.6%

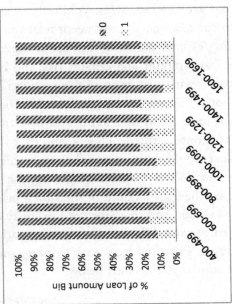

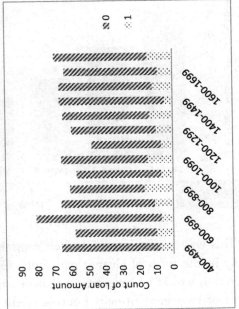

Though it would not have much importance for growth-stage investors, we also measured FPDs internally against the distribution of our average loan sizes. While this was a quick way to gain insight into our customers' credit needs relative to our underwriting and marketing efforts, we could also see if there was a relationship between the loan amount and default risk and begin looking for behavioral patterns with customers in the specified loan bucket.

For investors, *vintage* performance had to be stressed as the relevant measure of their return on investment: loans originated in the same month formed a cohort, and over the life span of the loans that cohort would return some (hopefully high) percentage of the capital lent out. When displayed together, the most recent vintages would recoup the least amount of their total principal, while the oldest vintages should have recouped the majority of theirs. This allowed us to organize capital deployment for originations by investor according to vintages. Principal recovery by vintage month was displayed to investors similar to the chart below:

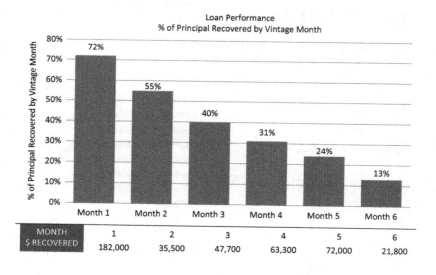

Loan Performance
% of Principal Recovered by Vintage Month

MONTH	1	2	3	4	5	6
$ RECOVERED	182,000	35,500	47,700	63,300	72,000	21,800

This was the rate at which their capital was returning to the company in the form of cleared payments. Over time, operational optimization would be reflected in mature vintages with higher average rates of recovered principal across credit cycles. The challenge was correctly formatting the data for this analysis: it was easy to track originations by month as each loan had a funding date, but a loan originated in September would have payments through April, so each payment needed to be identified as belonging to that loan and the sum of those (successful) payments had to be expressed as a percent of the total expected payments for the vintage month. Internally, we tracked vintage performance as curves that were refreshed according to processed payment updates, displayed in a graph similar to the following:

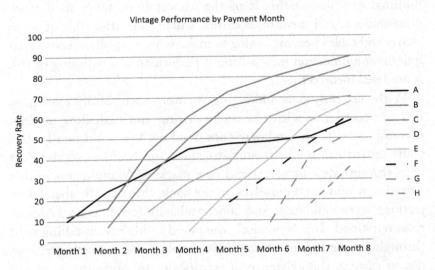

Though the curves are visually more appealing than the bars, we knew the chart as a whole was not as intuitive from an investor's perspective. The total principal recovered is lost as the eye moves left to right, as well as an efficient performance comparison between vintages. From an operations standpoint, however, management could focus on the recovery rates *over time*, highlighting performance trends over the life span of a vintage in relation to financial breakeven points. As we completed multiple vintage cycles quarter-over-quarter, we built a record of seasonal performance that would inform our capital raising and deployment schedules. Our historical performance helped predict the return our investors could expect.

The Problem with Analysts: Black Swans, ML, and the Future

While these examples highlight the effectiveness of elegant visualization and focused storytelling for lending data, audience-focused analytics communication applies to any business. Yet analysts by themselves are prone to throwing in as much information as a chart will support, including as many variables, indices, and moving averages as they can in an attempt to provide a "complete" picture of the data. This simply buries the highlights and obfuscates decision-making for audiences who are not working with the raw data every day. Use only the metrics that tell a meaningful, accurate story with minimal exposition. This is of the utmost importance in investor presentations, as I mentioned earlier, where slide after slide of dense charts and tables becomes a slog to make sense of. Still, effective analytics communication means little if performance is suffering or risk is modeled incorrectly.

With our early data indicating we were successfully managing risk in accordance with our underwriting and cash flow models, investors were happy and new ones lined up. We could finally start courting institutional investors to move the business to the next level through new products and new markets. Many teams in this situation would pat themselves on the back, think the financial projections were validated, and subconsciously feel like no systemic risks remained. This is precisely the mood I didn't want taking hold throughout the organization—it is simply naïve for entrepreneurs not to assume the existence of significant unknown risks, to lull themselves into thinking their model is stable because it performs in environments of "normal" volatility. The only way to identify our critical failures points[2] that could not be seen during day-to-day operations was to continually test the organization at its extremes. Analysts are often unwilling or unable to do this because statistically these events have extremely low probabilities of occurrence. Indeed, Nassim Taleb's infamous "black swan" events are so outside of statistical norms that they seem not just improbable, but impossible. Still, we developed an internal "Red Cell" team, based on the Navy Seals' 80s domestic counter-terrorism unit for testing U.S. security effectiveness, separate from our analysts to hack the organization's systems directly, both internally and externally. We also subjected the pro forma to sensitivity analysis, changing

---

[2] This includes system failures like IT outages and data hacks, but also payment processor issues, regulation and compliance changes, and sudden capital constraints affecting lending and operations.

key assumptions like the default rate then running and rerunning the model to determine performance thresholds. Analysts have a bad habit of supposing (or seeking) static assumptions, effectively "freezing" the pro forma such that successful performance becomes an end state rather than a process. The macroenvironment could suddenly shift the organization's benchmarks in a totally unrecognizable way, and when they finally realize it, their response is too late and the business is wiped out.

This latent volatility stresses the need for advanced data science and mature, robust data architecture to facilitate it. Organizations cannot eliminate risk entirely, nor can they truly predict black swans, but they can operate under the assumption that those events will occur. The big data paradigm shift has already occurred, yet many organizations with massive amounts of data (or data growth rapidly pushing them in that direction) are still only focused on the storage or engineering aspect rather than mining for insights. The data infrastructure discussed earlier was put in place precisely to leverage ML and automation: once our data lake is at scale, we can run ML algorithms to segment our customer base by expected profitability through feature selection of demographic and behavioral criteria, increase the accuracy of our underwriting and credit scoring models to further reduce risk, and analyze the unstructured call center data to improve our customer contact points. Through this, we can completely automate the underwriting, management, and collections processes for a seamless end-to-end customer experience.

The recursive optimization at the center of business intelligence today comes down to being able to grow an organization, not simply improving functional siloes. Indeed, the ideal data-centric organization is not built on siloes at all. Many organizations have very capable teams of analysts producing detailed reports that unfortunately have little relevance to the business's actual goals. This does not serve investors, management, or the analysts themselves. Rather, management must know and communicate their organization's outcomes with clarity, the analytical and operational teams must be able to measure and optimize towards those outcomes, and the data infrastructure must be constructed such that they can do so efficiently with all the tools of data science at their disposal. Only then will value be imparted—and actually returned—to all stakeholders.

"As You Can See ..."

SUBHASHISH SAMADDAR AND SATISH NARGUNDKAR

Georgia State University

Contents

How often have we all sat in on presentations where the speaker points to a projected slide and says something like "as you can see ...," but you could hardly "see" anything. This is typically because of the slide being crowded with text, data, and charts. When you show too much, the audience usually sees nothing. This is true for all audiences, even more so when the audience is made up of senior management who is pressed for time and has limited patience for extraneous detail.

For example, consider the presentation slide (Table 11.1) shown by a financial analyst to his CEO and other senior members of a company. The analyst put this slide on the screen and began by saying, "as you can see, for our major clients, when the discount is greater ...." It ensured that the audience was lost in the first sentence of the presentation! The table has 21 rows and 11 columns, for a total of 231 pieces of data. Aside from each number being barely visible, the audience simply did not know where to focus their attention.

The point the analyst wanted to make, which was presumably clear to him, was that there was an inverse relationship between the discounts offered to their large clients and the unit profit generated from those clients. This message is obscured by the forest of data in the table. The only information needed to convey that idea is the list of clients, the discounts offered, and the unit profits. In other words, only three columns out of the 11 presented were material to the point being made.

Table 11.1   Too Much Data

CLIENT NAME	QUANTITY	AVG PRICE	REVENUE	FIXED COST	VARIABLE COST	TOTAL COST	PROFIT	DISCOUNT	NET PROFIT	PROFIT/ UNIT
Client 1	880	2,460	2,164,800.00	1,000,000.00	880,000.00	1,880,000.00	284,800.00	20%	227,840.00	258.91
Client 2	913	2,460	2,245,980.00	1,000,000.00	913,000.00	1,913,000.00	332,980.00	13%	289,692.60	317.30
Client 3	535	2,470	1,321,450.00	307,000.00	535,000.00	842,000.00	479,450.00	5%	455,477.50	851.36
Client 4	856	2,200	1,883,200.00	400,000.00	856,000.00	1,256,000.00	627,200.00	10%	564,480.00	659.44
Client 5	110	2,500	275,000.00	100,000.00	110,000.00	210,000.00	65,000.00	6%	61,100.00	555.45
Client 6	650	2,450	1,592,500.00	330,000.00	650,000.00	980,000.00	612,500.00	6%	575,750.00	885.77
Client 7	151	2,500	377,500.00	100,000.00	151,000.00	251,000.00	126,500.00	3%	122,705.00	812.62
Client 8	522	2,460	1,284,120.00	304,400.00	522,000.00	826,400.00	457,720.00	6%	430,256.80	824.25
Client 9	105	2,500	262,500.00	50,000.00	105,000.00	155,000.00	107,500.00	1%	106,425.00	1,013.57
Client 10	446	2,480	1,106,080.00	289,200.00	446,000.00	735,200.00	370,880.00	5%	352,336.00	789.99
Client 11	710	2,450	1,739,500.00	600,000.00	710,000.00	1,310,000.00	429,500.00	10%	386,550.00	544.44
Client 12	250	2,500	625,000.00	250,000.00	250,000.00	500,000.00	125,000.00	9%	113,750.00	455.00
Client 13	635	2,460	1,562,100.00	327,000.00	635,000.00	962,000.00	600,100.00	7%	558,093.00	878.89
Client 14	363	2,480	900,240.00	272,600.00	363,000.00	635,600.00	264,640.00	5%	251,408.00	692.58
Client 15	511	2,450	1,251,950.00	302,200.00	511,000.00	813,200.00	438,750.00	5%	416,812.50	815.68
Client 16	379	2,460	932,340.00	275,800.00	379,000.00	654,800.00	277,540.00	4%	266,438.40	703.00
Client 17	884	2,400	2,121,600.00	800,000.00	884,000.00	1,684,000.00	437,600.00	15%	371,960.00	420.77
Client 18	310	2,470	765,700.00	200,000.00	310,000.00	510,000.00	255,700.00	4%	245,472.00	791.85
Client 19	281	2,475	695,475.00	200,000.00	281,000.00	481,000.00	214,475.00	6%	201,606.50	717.46
Client 20	171	2,500	427,500.00	100,000.00	171,000.00	271,000.00	156,500.00	2%	153,370.00	896.90

A better way to show the table would therefore have been to restrict the data to those three columns. This would clear the forest of all the distracting columns, and only the relevant data would be shown (Table 11.2).

The relationship between discounts and unit profit is still not obvious from the table. The data are organized alphabetically by client name, which does nothing to show the relationship between discounts and profit. The next level of improvement would be to sort the columns in either ascending or descending order by the discount provided, to see if a casual inspection would reveal a pattern (Table 11.3).

This table now permits a rough conclusion that there seems to be an inverse relationship between the variables shown. It is still difficult to process 40 numbers in one's head. To truly see the pattern, a simple scatterplot will do the job (Figure 11.1).

Given how simple it is to show the scatterplot to make the key point, why do analysts often show the entire data as shown in Table 11.1? The answer probably is that the analyst has spent many

Table 11.2 Relevant Data

CLIENT NAME	DISCOUNT (%)	PROFIT/UNIT
Client 1	20	258.91
Client 2	13	317.3
Client 3	5	851.36
Client 4	10	659.44
Client 5	6	555.45
Client 6	6	885.77
Client 7	3	812.62
Client 8	6	824.25
Client 9	1	1,013.57
Client 10	5	789.99
Client 11	10	544.44
Client 12	9	455
Client 13	7	878.89
Client 14	5	692.58
Client 15	5	815.68
Client 16	4	703
Client 17	15	420.77
Client 18	4	791.85
Client 19	6	717.46
Client 20	2	896.9

Table 11.3   Data Sorted by Discount

CLIENT NAME	DISCOUNT (%)	PROFIT/UNIT
Client 1	20	258.91
Client 17	15	420.77
Client 2	13	317.30
Client 4	10	659.44
Client 11	10	544.44
Client 12	9	455.00
Client 13	7	878.89
Client 5	6	555.45
Client 6	6	885.77
Client 8	6	824.25
Client 19	6	717.46
Client 3	5	851.36
Client 10	5	789.99
Client 14	5	692.58
Client 15	5	815.68
Client 16	4	703.00
Client 18	4	791.85
Client 7	3	812.62
Client 20	2	896.90
Client 9	1	1013.57

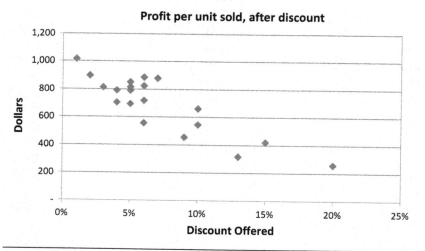

Figure 11.1   Unit profit versus discount.

hours compiling the data and making several computations before reaching the key insights, and feels compelled to share all the intermediate results. The presumption perhaps is that this will impress the audience with how much hard work went into this process and might help with audience buy-in.

A graph can be a great way to summarize multiple numbers and help relationships stand out. However, charts and graphs can be over-crowded too, and defeat the purpose. We attended a presentation by a retired, high-ranking government official talking about the economies of South American countries. He put up a chart shown in Figure 11.2 and said something like "As you can see, the GDP growth of Cuba is ... ." Someone from the audience did say, quite politely, that he could not see on the chart what was being said.

The chart is informative, but highly cluttered, thus making it difficult for the audience follow the speaker's point. Assuming that the point is to discuss Cuba's growth rate and compare it with the growth rate of other countries, the chart can be decluttered. One way would be to show a chart with just Cuba's growth rate first, and then show a chart including Cuba's growth rate and perhaps the average growth rate of all the others (or the growth rate of a single representative country among the others).

Finally, in our own work, we came across a similar situation. A client sent us data to analyze the variability in loading times at various locations for shipments bound for a particular destination. The client asked for a model to predict the loading times based on location and other variables. Their true goal, not disclosed to us at the time, was to identify the cause of certain anomalies in the loading times.

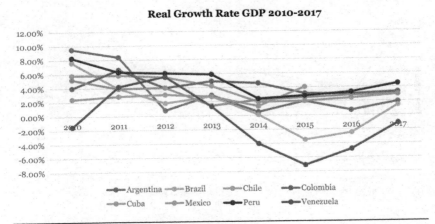

Figure 11.2 GDP growth of South American Nations.

They were unhappy with the overall performance of their company in terms of loading times and delays and did not have a complete understanding of the causes.

Our initial descriptive analysis led us to the chart shown in Figure 11.3.

The chart looked too cluttered, so we decided to keep only the lines that showed a marked increase or decrease over time. The ones that remained steady were uninteresting from a decision making standpoint. Management would be interested in knowing why loading times were going up, indicating a problem to be solved, or going down, indicating efficiencies achieved that could be emulated elsewhere. This led us to the chart shown in Figure 11.4.

Once the graph was narrowed down to these four lines, it was clear that two were going up over time and the other two going down. For the purposes of presentation to the client, we decided to split this further into two charts, as shown in Figures 11.5 and 11.6.

As soon as the clients saw these two charts, they told us that they now knew what the cause of the variations was. At this point, we had no idea what the cause was. Later, they told us that the charts clearly revealed to them the months and locations during which there were drastic changes. That was the key that led to a sudden flash of insight in their minds, given the knowledge only they had about events that transpired during those critical months (we are withholding that

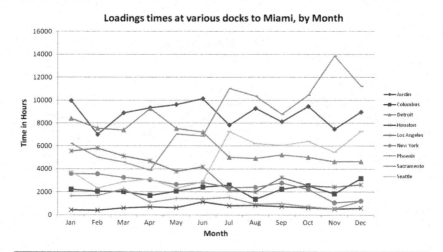

Figure 11.3   Loading times, cluttered.

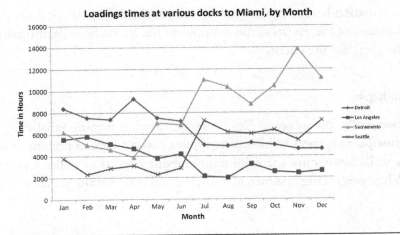

Figure 11.4 Loading times with most variation.

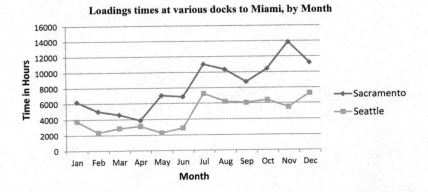

Figure 11.5 Increasing loading times.

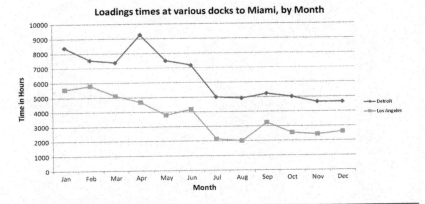

Figure 11.6 Decreasing loading times.

information here to protect the client's anonymity). Our not knowing the cause of the variation did not prevent the charts from illuminating the cause for the client.

Epilogue

Essentially, the adage of "less is more," used as a minimalist design principle in art and architecture, applies to the design of presentations as well. Show only the minimum amount needed to make a point. When everything is shown at once, the audience usually sees nothing.

Index